Modern Approaches to Advanced Calculus

Modern Approaches to Advanced Calculus

Rakesh Goyal

Oxford Book Company
Jaipur ▪ New Delhi

Published By

Oxford Book Company

Regd. Off. "Bony Residency", Gate No. 2
Opp. Tilak Public School, Vishweswariya Nagar,
Gopalpura Road, Jaipur - 302018, India
Ph: 0141 - 2761280, Telefax: 2761381
email - oxfordbookjpr@gmail.com
website - oxfordbookcompany.com

Delhi Office: 102, 1st Floor, Satyam House,
4327/3, Ansari Road, Darya Ganj,
New Delhi - 110002, India
Ph: 011-45652440

Distributed By
KEYA INTERNATIONAL PRESS LLP
E-mail: keyainternationalpressllp@gmail.com

First published in India, 2026

ISBN : 978-93-5524-322-5

Typeset by Shivangi Computers, Jaipur

Printed and bound in India by Thomson Press Ltd.,
New Delhi

Contents

Preface

Advanced calculus stands as the cornerstone of mathematical analysis, offering profound insights into the nature of continuous change, limits, differentiation, and integration. Its applications extend across disciplines such as physics, engineering, economics, and computer science, making it an indispensable tool for scientific discovery and technological advancement. Modern Approaches to Advanced Calculus is designed to provide a rigorous yet accessible exploration of advanced calculus, integrating classical concepts with contemporary perspectives to cater to the evolving needs of students, educators, and researchers.

This book offers a structured approach to understanding key topics in advanced calculus, beginning with foundational principles such as real analysis, sequences, and series, before advancing to multivariable calculus, vector analysis, and differential equations. Emphasis is placed on the logical development of mathematical concepts, ensuring that readers not only grasp computational techniques but also develop a deeper appreciation for theoretical underpinnings. Additionally, the book explores emerging areas such as fractional calculus, variational principles, and the interplay between calculus and modern computational methods.

A distinguishing feature of Modern Approaches to Advanced Calculus is its balance between abstraction and application. Each concept is presented with mathematical rigor while being reinforced with real-world examples, visual interpretations, and problem-solving exercises. The inclusion of historical insights provides context on the evolution of calculus, tracing its development from the pioneering works of Newton and Leibniz to its modern-day applications in complex analysis, optimization, and artificial intelligence.

To further enhance comprehension, this book integrates technology-driven approaches, demonstrating the use of software tools such as MATLAB, Mathematica, and Python for solving advanced calculus problems. These computational techniques not only aid in visualization but also offer practical problem-solving strategies, making calculus more intuitive and applicable in scientific research and industry applications.

The current book was developed with undergraduate and graduate students at various Indian universities in mind. The subject matter, both in terms of chapter organisation and substance, has been graded so that students may follow the course with ease and enjoyment, moving from easy to more complex themes.

— ***Rakesh Goyal***

Chapter 1

Introduction

The fundamental theorem of calculus is a theorem that links the concept of the derivative of a function with the concept of the integral. The first part of the theorem, sometimes called the first fundamental theorem of calculus, shows that an indefinite integration can be reversed by a differentiation. This part of the theorem is also important because it guarantees the existence of antiderivatives for continuous functions. The second part, sometimes called the second fundamental theorem of calculus, allows one to compute the definite integral of a function by using any one of its infinitely many antiderivatives. This part of the theorem has invaluable practical applications, because it markedly simplifies the computation of definite integrals. The first published statement and proof of a restricted version of the fundamental theorem was by James Gregory (1638–1675). Isaac Barrow (1630–1677) proved a more generalized version of the theorem, while Barrow's student Isaac Newton (1643–1727) completed the development of the surrounding mathematical theory. Gottfried Leibniz (1646–1716) systematized the knowledge into a calculus for infinitesimal quantities, and introduced the notation used today.

Physical Intuition

Intuitively, the theorem simply states that the sum of infinitesimal changes in a quantity over time (or over some other quantity) adds up to the net change in the quantity. In the case of a particle travelling in a straight line, its position, x, is given by $x(t)$ where t is time and $x(t)$ means that x is a function of t. The derivative of this function is equal to the infinitesimal change in quantity, dx, per infinitesimal change in time, dt (of course, the derivative itself is dependent on time). This change in displacement per change in time is the velocity v of the particle. In Leibniz's notation:

$$\frac{dx}{dt} = v(t).$$

Rearranging this equation, it follows that:

$$dx = v(t)dt.$$

By the logic above, a change in x (or Δx) is the sum of the infini-tesimal changes dx. It is also equal to the sum of the infinitesimal products of the derivative and time. This infinite summation is integration; hence, the integration operation allows the recovery of the original function from its derivative. It can be concluded that this operation works in reverse; the result of the integral can be differentiated to recover the original function.

Geometric Intuition

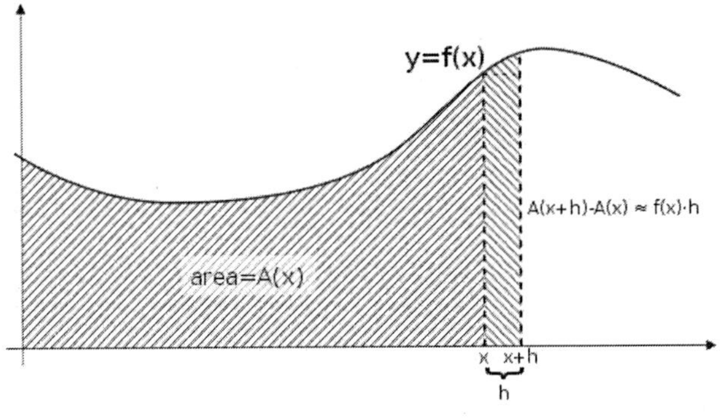

Figure: *The area shaded in red stripes can be estimated as* h *times* f(x). *Alternatively, if the function* A(x) *were known, it could be computed as* A(x + h) - A(x). *These two values are approximately equal, particularly for small* h.

For a continuous function $y = f(x)$ whose graph is plotted as a curve, each value of x has a corresponding area function $A(x)$, representing the area beneath the curve between 0 and x. The function $A(x)$ may not be known, but it is given that it represents the area under the curve. The area under the curve between x and $x + h$ could be computed by finding the area between 0 and $x + h$, then subtracting the area between 0 and x. In other words, the area of this "sliver" would be $A(x + h) - A(x)$.

There is another way to *estimate* the area of this same sliver. h is multiplied by $f(x)$ to find the area of a rectangle that is approximately the same size as this sliver.

It is intuitive that the approximation improves as h becomes smaller.

At this point, it is true $A(x + h) - A(x)$ is approximately equal to $f(x) \cdot h$. In other words,

$$f(x)h \approx A(x+h) - A(x),$$

with this approximation becoming an equality as h approaches 0 in the limit.

When both sides of the equation are divided by h:

$$f(x) \approx \frac{A(x+h) - A(x)}{h}.$$

As h approaches 0, it can be seen that the right hand side of this equation is simply the derivative $A2\ (x)$ of the area function $A(x)$. The left-hand side of the equation simply remains $f(x)$, since no h is present.

It can thus be shown, in an informal way, that $f(x) = A2\ (x)$. That is, the derivative of the area function $A(x)$ is the original function $f(x)$; or, the area function is simply the antiderivative of the original function. Computing the derivative of a function and "finding the area" under its curve are "opposite" operations. This is the crux of the Fundamental Theorem of Calculus. Most of the theorem's proof is devoted to showing that the area function $A(x)$ exists in the first place, under the right conditions.

Formal Statements

There are two parts to the Fundamental Theorem of Calculus. Loosely put, the first part deals with the derivative of an antiderivative, while the second part deals with the relationship between antideri-vatives and definite integrals.

First Part

This part is sometimes referred to as the First Fundamental Theorem of Calculus.

Let f be a continuous real-valued function defined on a closed interval $[a, b]$. Let F be the function defined, for all x in $[a, b]$, by

$$F(x) = \int_a^x f(t)dt.$$

Then, F is continuous on $[a, b]$, differentiable on the open interval (a, b), and

$$F'(x) = f(x)$$

for all x in (a, b).

Corollary

The fundamental theorem is often employed to compute the definite integral of a function f for which an antiderivative g is known. Specifically, if f is a real-valued continuous function on $[a, b]$, and F is an antiderivative of f in $[a, b]$, then

$$\int_a^b f(x)dx = F(b) - F(a).$$

The corollary assumes continuity on the whole interval. This result is strengthened slightly in the following part of the theorem.

Second Part

This part is sometimes referred to as the Second Fundamental Theorem of Calculus or the Newton–Leibniz Axiom.

Let f be a real-valued function defined on a closed interval $[a, b]$ that admits an antiderivative g on $[a, b]$. That is, f and g are functions such that for all x in $[a, b]$,

$$f(x) = g'(x).$$

If f is Riemann integrable on $[a, b]$ then

$$\int_a^b f(x)dx = g(b) - g(a).$$

Notice that the Second part is somewhat stronger than the Corollary because it does not assume that f is continuous. Note that when an antiderivative g exists, then there are infinitely many antiderivatives for f, obtained by adding to g an arbitrary constant.

Also, by the first part of the theorem, antiderivatives of f always exist when f is continuous.

Proof of the First Part

For a given $f(t)$, define the function $F(x)$ as

$$F(x) = \int_a^x f(t)dt.$$

For any two numbers x_1 and $x_1 + \Delta x$ in $[a, b]$, we have

$$F(x_1) = \int_a^{x_1} f(t)dt$$

and

$$F(x_1 + \Delta x) = \int_a^{x_1+\Delta x} f(t)dt.$$

Subtracting the two equations gives

$$F(x_1 + \Delta x) - F(x_1) = \int_a^{x_1+\Delta x} f(t)dt - \int_a^{x_1} f(t)dt. \qquad (1)$$

It can be shown that

$$\int_a^{x_1} f(t)dt + \int_{x_1}^{x_1+\Delta x} f(t)dt = \int_a^{x_1+\Delta x} f(t)dt.$$

(The sum of the areas of two adjacent regions is equal to the area of both regions combined.)

Manipulating this equation gives

$$\int_a^{x_1+\Delta x} f(t)dt - \int_a^{x_1} f(t)dt = \int_{x_1}^{x_1+\Delta x} f(t)dt.$$

Substituting the above into (1) results in

$$F(x_1 + \Delta x) - F(x_1) = \int_{x_1}^{x_1+\Delta x} f(t)dt. \qquad (2)$$

According to the mean value theorem for integration, there exists a c in $[x_1, x_1 + \Delta x]$ such that

$$\int_{x_1}^{x_1+\Delta x} f(t)dt = f(c)\Delta x.$$

Substituting the above into (2) we get

$$F(x_1 + \Delta x) - F(x_1) = f(c)\Delta x.$$

Dividing both sides by Δx gives

$$\frac{F(x_1 + \Delta x) - F(x_1)}{\Delta x} = f(c).$$

Notice that the expression on the left side of the equation is Newton's difference quotient for F at x_1.

Take the limit as $\Delta x \to 0$ on both sides of the equation.

$$\lim_{\Delta x \to 0} \frac{F(x_1 + \Delta x) - F(x_1)}{\Delta x} = \lim_{\Delta x \to 0} f(c).$$

The expression on the left side of the equation is the definition of the derivative of F at x_1.

$$F'(x_1) = \lim_{\Delta x \to 0} f(c). \qquad (3)$$

To find the other limit, we will use the squeeze theorem. The number c is in the interval $[x_1, x_1 + \Delta x]$, so $x_1 \le c \le x_1 + \Delta x$.

Also, $\lim_{\Delta x \to 0} x_1 = x_1$ and $\lim_{\Delta x \to 0} x_1 + \Delta x = x_1$.

Therefore, according to the squeeze theorem,

$$\lim_{\Delta x \to 0} c = x_1.$$

Substituting into (3), we get

$$F'(x_1) = \lim_{c \to x_1} f(c).$$

The function f is continuous at c, so the limit can be taken inside the function.

Therefore, we get

$$F'(x_1) = f(x_1).$$

which completes the proof.

(Leithold *et al.*, 1996)

Proof of the Corollary

Let $F(x)=\int_a^x f(t)dt$, with f continuous on $[a, b]$. If g is an antiderivative of f, then g and F have the same derivative, by the *first part* of the theorem. It follows by the mean value theorem that there is a number c such that $F(x) = g(x) + c$, for all x in $[a, b]$. Letting $x = a$,

$$F(a)=\int_a^a f(t)dt=0=g(a)+c,$$

which m eans $c = -g(a)$. In other words $F(x) = g(x) - g(a)$, and so

$$\int_a^b f(x)dx=g(b)-g(a).$$

Proof of the Second Part

This is a limit proof by Riemann sums. Let f be (Riemann) integrable on the interval $[a, b]$, and let f admit an antiderivative F on $[a, b]$. Begin with the quantity $F(b) - F(a)$. Let there be numbers $x_1, ..., x_n$ such that

$$a=x_0<x_1<x_2<\cdots<x_{n-1}<x_n=b.$$

It follows that

$$F(b)-F(a)=F(x_n)-F(x_0).$$

Now, we add each $F(x_i)$ along with its additive inverse, so that the resulting quantity is equal:

$$\begin{aligned} F(b)-F(a) &= F(x_n)+[-F(x_{n-1})+F(x_{n-1})]+\cdots+[-F(x_1)+F(x_1)]-F(x_0) \\ &= [F(x_n)-F(x_{n-1})]+[F(x_{n-1})+\cdots-F(x_1)]+(F(x_1)-F(x_0)). \end{aligned}$$

The above quantity can be written as the following sum:

$$F(b)-F(a)=\sum_{i=1}^{n}[F(x_i)-F(x_{i-1})]. \qquad (1)$$

Next we will employ the mean value theorem. Stated briefly,

Let F be continuous on the closed interval $[a, b]$ and differentiable on the open interval (a, b). Then there exists some c in (a, b) such that

$$F'(c) = \frac{F(b) - F(a)}{b - a}.$$

It follows that

$$F'(c)(b - a) = F(b) - F(a).$$

The function F is differentiable on the interval $[a, b]$; therefore, it is also differentiable and continuous on each interval $[x_{i-1}, x_i]$. According to the mean value theorem (above),

$$F(x_i) - F(x_{i-1}) = F'(c_i)(x_i - x_{i-1}).$$

Substituting the above into (1), we get

$$F(b) - F(a) = \sum_{i=1}^{n} [F'(c_i)(x_i - x_{i-1})].$$

The assumption implies $F'(c_i) = f(c_i)$. Also, $x_i - x_{i-1}$ can be expressed as Δx of partition i.

$$F(b) - F(a) = \sum_{i=1}^{n} [f(c_i)(\Delta x_i)]. \qquad (2)$$

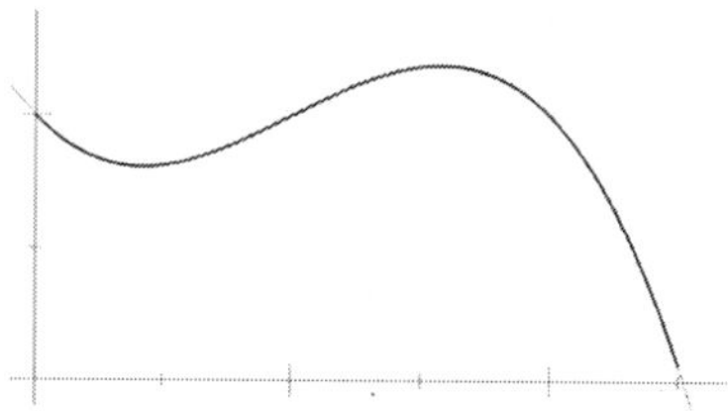

Figure: *A converging sequence of Riemann sums. The number in the upper left is the total area of the blue rectangles. They converge to the integral of the function.*

Notice that we are describing the area of a rectangle, with the width times the height, and we are adding the areas together. Each rectangle, by virtue of the Mean Value Theorem, describes an approximation of the curve section it is drawn over. Also notice that Δx_i need not be the same for all values of i, or in other words that the width of the rectangles can differ. What we have to do is approximate the curve with n rectangles. Now, as the size of the partitions get smaller and n increases, resulting in more partitions to cover the space, we will get closer and closer to the actual area of the curve.

By taking the limit of the expression as the norm of the partitions approaches zero, we arrive at the Riemann integral. We know that this limit exists because f was assumed to be integrable. That is, we take the limit as the largest of the partitions approaches zero in size, so that all other partitions are smaller and the number of partitions approaches infinity.

So, we take the limit on both sides of (2). This gives us

$$\lim_{\|\Delta\|\to 0} F(b) - F(a) = \lim_{\|\Delta\|\to 0} \sum_{i=1}^{n} [f(c_i)(\Delta x_i)].$$

Neither $F(b)$ nor $F(a)$ is dependent on $||\Delta||$, so the limit on the left side remains $F(b)$ " $F(a)$.

$$F(b) - F(a) = \lim_{\|\Delta\|\to 0} \sum_{i=1}^{n} [f(c_i)(\Delta x_i)].$$

The expression on the right side of the equation defines the integral over f from a to b. Therefore, we obtain

$$F(b) - F(a) = \int_a^b f(x)dx,$$

which completes the proof.

It almost looks like the first part of the theorem follows directly from the second, because the equation $g(x) - g(a) = \int_a^x f(t)dt$, where g is an antiderivative of f, implies that $F(x) = \int_a^x f(t)dt$ has the same derivative as g, and therefore $F2 = f$. This argument only works if we already know that f has an antiderivative, and the only way we know that all continuous functions have antiderivatives is by the first part of the Fundamental Theorem. For example if $f(x) = e^{-x2}$, then f has an antiderivative, namely

$$g(x) = \int_0^x f(t)dt$$

and there is no simpler expression for this function. It is therefore important not to interpret the second part of the theorem as the definition of the integral. Indeed, there are many functions that are integrable but lack antiderivatives that can be written as an elementary function. Conversely, many functions that have antiderivatives are not Riemann integrable.

Generalizations

We don't need to assume continuity of f on the whole interval. Part I of the theorem then says: if f is any Lebesgue integrable function on $[a, b]$ and x_0 is a number in $[a, b]$ such that f is continuous at x_0, then

$$F(x) = \int_a^x f(t)dt$$

is differentiable for $x = x_0$ with $F2\ (x_0) = f(x_0)$. We can relax the conditions on f still further and suppose that it is merely locally integrable. In that case, we can conclude that the function F is differentiable almost everywhere and $F2\ (x) = f(x)$ almost everywhere. On the real line this statement is equivalent to Lebesgue's differen-tiation theorem. These results remain true for the Henstock–Kurzweil integral which allows a larger class of integrable functions (Bartle 2001, Thm. 4.11).

In higher dimensions Lebesgue's differentiation theorem generalizes the Fundamental theorem of calculus by stating that for almost every x, the average value of a function f over a ball of radius r centered at x will tend to $f(x)$ as r tends to 0.

Part II of the theorem is true for any Lebesgue integrable function f which has an antiderivative F (not all integrable functions do, though). In other words, if a real function F on $[a, b]$ admits a derivative $f(x)$ at *every* point x of $[a, b]$ and if this derivative f is Lebesgue integrable on $[a, b]$, then

$$F(b) - F(a) = \int_a^b f(t)dt.$$

This result may fail for continuous functions F that admit a derivative $f(x)$ at almost every point x, as the example of the Cantor function shows. But the result remains true if F is absolutely continuous: in that case, F admits a derivative $f(x)$

at almost every point x and, as in the formula above, $F(b)$ ™ $F(a)$ is equal to the integral of f on $[a, b]$. The conditions of this theorem may again be relaxed by considering the integrals involved as Henstock–Kurzweil integrals. Specifically, if a continuous function $F(x)$ admits a derivative $f(x)$ at all but countably many points, then $f(x)$ is Henstock–Kurzweil integrable and $F(b)$ ™ $F(a)$ is equal to the integral of f on $[a, b]$. The difference here is that the integrability of f does not need to be assumed. (Bartle 2001, Thm. 4.7) The version of Taylor's theorem which expresses the error term as an integral can be seen as a generalization of the Fundamental Theorem. There is a version of the theorem for complex functions: suppose U is an open set in C and $f: U \to$ C is a function which has a holomorphic antiderivative F on U. Then for every curve γ : $[a, b] \to U$, the curve integral can be computed as

$$\int_{\gamma} f(z)dz = F(\gamma(b)) - F(\gamma(a)).$$

The fundamental theorem can be generalized to curve and surface integrals in higher dimensions and on manifolds. One such generali-zation offered by the calculus of moving surfaces is the time evolution of integrals. The most familiar extensions of the Fundamental theorem of calculus in two dimensions are Green's theorem and the two-dimen-sional case of the Gradient theorem. One of the most powerful statements in this direction is Stokes' theorem: Let M be an oriented piecewise smooth manifold of dimension n and let ω be an n-1 form that is a compactly supported differential form on M of class C^1. If ∂M denotes the boundary of M with its induced orientation, then

$$\int_{M} d\omega = \oint_{\partial M} \omega.$$

Here d is the exterior derivative, which is defined using the manifold structure only. The theorem is often used in situations where M is an embedded oriented submanifold of some bigger manifold on which the form ω is defined.

Rolle's Theorem

In calculus, Rolle's theorem essentially states that a differentiable function which attains equal values at two distinct points must have a point somewhere between them where the first derivative (the slope of the tangent line to the graph of the function) is zero.

Standard Version of the Theorem

If a real-valued function f is continuous on a closed interval $[a, b]$, differentiable on the open interval (a, b), and $f(a) = f(b)$, then there exists a c in the open interval (a, b) such that

$$f'(c) = 0.$$

This version of Rolle's theorem is used to prove the mean value theorem, of which Rolle's theorem is indeed a special case. It is also the basis for the proof of the Taylor's theorem.

History

The first known formal proof was offered by Michel Rolle in 1691, which used the methods of differential calculus. The name "Rolle's theorem" was first used by Moritz Wilhelm Drobisch of Germany in 1834 and by Giusto Bellavitis of Italy in 1846.

Examples

First example:

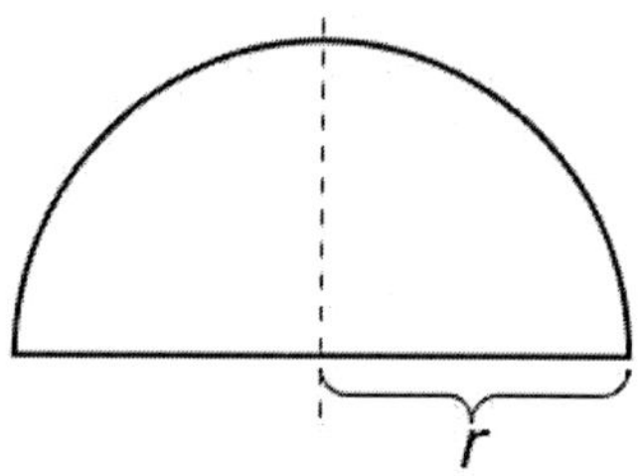

Figure: *A semicircle of radius* r

For a radius $r > 0$ consider the function

$$f(x) = \sqrt{r^2 - x^2}, \quad x \in [-r, r].$$

Its graph is the upper semicircle centered at the origin. This function is continuous on the closed interval $[-r,r]$ and differentiable in the open interval $(-r,r)$, but not differentiable at the endpoints $-r$ and r. Since $f(-r) = f(r)$, Rolle's theorem applies, and indeed, there is a point where the derivative of f

is zero. Note that the theorem applies even when the function cannot be differentiated at the endpoints because it only requires the function to be differentiable in the open interval.

Second Example

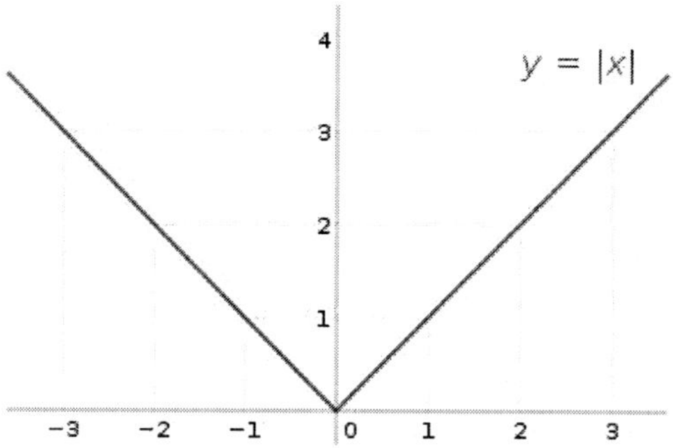

Figure: *The graph of the absolute value function*

If differentiability fails at an interior point of the interval, the conclusion of Rolle's theorem may not hold. Consider the absolute value function

$$f(x) = |x|, \qquad x \in [-1, 1].$$

Then $f(-1) = f(1)$, but there is no c between -1 and 1 for which the derivative is zero. This is because that function, although continuous, is not differentiable at $x = 0$. Note that the derivative of f changes its sign at $x = 0$, but without attaining the value 0. The theorem cannot be applied to this function, clearly, because it does not satisfy the condition that the function must be differentiable for every x in the open interval.

Generalization

The second example illustrates the following generalization of Rolle's theorem:

Consider a real-valued, continuous function f on a closed interval $[a,b]$ with $f(a) = f(b)$. If for every x in the open interval (a,b) the right-hand limit

$$f'(x+) := \lim_{h \to 0^+} \frac{f(x+h) - f(x)}{h}$$

and the left-hand limit

$$f'(x-) := \lim_{h \to 0^-} \frac{f(x+h) - f(x)}{h}$$

exist in the extended real line $[-\infty,\infty]$, then there is some number c in the open interval (a,b) such that one of the two limits

$$f'(c+) \quad \text{and} \quad f'(c-)$$

is ≥ 0 and the other one is ≤ 0 (in the extended real line). If the right- and left-hand limit agree for every x, then they agree in particular for c, hence the derivative of f exists at c and is equal to zero.

Remarks

1. If f is convex or concave, then the right- and left-hand derivatives exist at every inner point, hence the above limits exist and are real numbers.
2. This generalized version of the theorem is sufficient to prove convexity when the one-sided derivatives are monotonically increasing:

$$f'(x-) \leq f'(x+) \leq f'(y-), \qquad x < y.$$

Proof of the Generalized Version

Since the proof for the standard version of Rolle's theorem and the generalization are very similar, we prove the generalization.

The idea of the proof is to argue that if $f(a) = f(b)$, then f must attain either a maximum or a minimum somewhere between a and b, say at c, and the function must change from increasing to decreasing (or the other way around) at c.

In particular, if the derivative exists, it must be zero at c. By assumption, f is continuous on $[a,b]$, and by the extreme value theorem attains both its maximum and its minimum in $[a,b]$.

If these are both attained at the endpoints of $[a,b]$, then f is constant on $[a,b]$ and so the derivative of f is zero at every point in (a,b).

Suppose then that the maximum is obtained at an interior point c of (a,b) (the argument for the minimum is very similar, just consider $-f$). We shall examine the above right- and left-hand limits separately. For a real h such that $c + h$ is in $[a,b]$, the value $f(c + h)$ is smaller or equal to $f(c)$ because f attains its maximum at c. Therefore, for every $h > 0$,

$$\frac{f(c+h)-f(c)}{h} \leq 0,$$

hence

$$f'(c+) := \lim_{h\to 0^+} \frac{f(c+h)-f(c)}{h} \leq 0,$$

where the limit exists by assumption, it may be minus infinity.

Similarly, for every $h < 0$, the inequality turns around because the denominator is now negative and we get

$$\frac{f(c+h)-f(c)}{h} \geq 0,$$

hence

$$f'(c-) := \lim_{h\to 0^-} \frac{f(c+h)-f(c)}{h} \geq 0,$$

where the limit might be plus infinity.

Finally, when the above right- and left-hand limits agree (in particular when f is differentiable), then the derivative of f at c must be zero.

Generalization to Higher Derivatives

We can also generalize Rolle's theorem by requiring that f has more points with equal values and greater regularity. Specifically, suppose that

- the function f is n - 1 times continuously differentiable on the closed interval $[a,b]$ and the nth derivative exists on the open interval (a,b), and
- there are n intervals given by $a_1 < b_1$ d $\leq a_2 < b_2 \leq . . <$ $a_n < b_n$ in $[a,b]$ such that $f(a_k) = f(b_k)$ for every k from 1 to n.

Then there is a number c in (a,b) such that the nth derivative of f at c is zero. Of course, the requirements concerning the nth derivative of f can be weakened as in the generalization above, giving the corres-ponding (possibly weaker) assertions for the right- and left-hand limits defined above with $f^{(n-1)}$ in place of f.

Proof

The proof uses mathematical induction. For $n = 1$ is simply the standard version of Rolle's theorem. As induction hypothesis, assume the generalization is true for $n - 1$. We want to prove it for $n > 1$. By the standard version of Rolle's theorem, for every integer k from 1 to n, there exists a c_k in the open interval (a_k, b_k) such that $f'(c_k) = 0$. Hence the first derivative satisfies the assumptions with the $n - 1$ closed intervals $[c_1, c_2], \ldots, [c_{n-1}, c_n]$. By the induction hypothesis, there is a c such that the $(n - 1)^{st}$ derivative of f' at c is zero.

Generalizations to Other Fields

Rolle's theorem is a property of differentiable functions over the real numbers, which are an ordered field. As such, it does not generalize to other fields, but the following corollary does: if a real polynomial splits (has all its roots) over the real numbers, then its derivative does as well – one may call this property of a field Rolle's property.

More general fields do not always have a notion of differentiable function, but they do have a notion of polynomials, which can be symbolically differentiated. Similarly, more general fields may not have an order, but one has a notion of a root of a polynomial lying in a field.

Thus Rolle's theorem shows that the real numbers have Rolle's property, and any algebraically closed field such as the complex numbers has Rolle's property, but conversely the rational numbers do not – for example, $x^3 - x = x(x-1)(x+1)$ splits over the rationals, but its derivative $3x^2 - 1 = 3(x - 1/\sqrt{3})(x + 1/\sqrt{3})$ does not. The question of which fields satisfy Rolle's property was raised in (Kaplansky 1972). For finite fields, the answer is that only $\mathbf{F}_2$ and $\mathbf{F}_4$ have Rolle's property; this was first proven via technical means in (Craven & Csordas 1977), and a simple proof is given in (Ballantine & Roberts 2002).

Quantum Calculus

Quantum calculus, sometimes called calculus without limits, is equivalent to traditional infinitesimal calculus without the notion of limits. It defines "q-calculus" and "ħ-calculus". h ostensibly stands for Planck's constant while q stands for quantum. The two parameters are related by the formula

$$q = e^{ih} = e^{2\pi i\hbar}$$

where $\hbar = \dfrac{h}{2\pi}$ is the reduced Planck constant.

Differentiation

In the q-calculus and h-calculus, differentials of functions are defined as

$$d_h(f(x)) = f(x+h) - f(x)$$

and

$$d_h(f(x)) = f(x+h) - f(x)$$

respectively. Derivatives of functions are then defined as fractions by the q-derivative

$$D_q(f(x)) = \frac{d_q(f(x))}{d_q(x)} = \frac{f(qx) - f(x)}{(q-1)x}$$

and by

$$D_h(f(x)) = \frac{d_h(f(x))}{d_h(x)} = \frac{f(x+h) - f(x)}{h}$$

In the limit, as h goes to 0, or equivalently as q goes to 1, these expressions take on the form of the derivative of classical calculus.

Integration

q-integral: A function F(x) is a q-antiderivative of f(x) if $D_q F(x) = f(x)$. The q-antiderivative (or q-integral) is denoted by $\int f(x) d_q x$ and an expression for F(x) can be found from the formula $\int f(x) d_q x = (1-q)\sum_{j=0}^{\infty} xq^J f(xq^j)$ which is called the Jackson integral of f(x). For 0 < q < 1, the series converges to a function F(x) on an interval (0,A] if |f(x)x^α| is bounded on the interval (0,A] for some 0 <= α < 1.

The q-integral is a Riemann-Stieltjes integral with respect to a step function having infinitely many points of increase at the points q^j, with the jump at the point q^j being q^j. If we call this step function $g_q(t)$ then $dg_q(t) = d_q t$.

h-integral

A function F(x) is an h-antiderivative of f(x) if $D_h F(x)=f(x)$. The h-antiderivative (or h-integral) is denoted by $\int f(x)d_h x$. If a and b differ by an integer multiple of h then the definite integral $\int_a^b f(x)d_h x$ is given by a Riemann sum of f(x) on the interval [a,b] partitioned into subintervals of width h.

Example

The derivative of the function x^n (for some positive integer n) in the classical calculus is nx^{n-1}. The corresponding expressions in q-calculus and h-calculus are

$$D_q(x^n) = \frac{q^n - 1}{q - 1} x^{n-1} = [n]_q \, x^{n-1}$$

with the q-bracket

$$[n]_q = \frac{q^n - 1}{q - 1}$$

and

$$D_h(x^n) = x^{n-1} + hx^{n-2} + \cdots + h^{n-1}$$

respectively. The expression $[n]_q x^{n-1}$ is then the q-calculus analogue of the simple power rule for positive integral powers. In this sense, the function x_n is still *nice* in the q-calculus, but rather ugly in the h-calculus – the h-calculus analog of x_n is instead the falling factorial $(x)n := x(x-1)\ldots(x - n + 1)$.

One may proceed further and develop, for example, equivalent notions of Taylor expansion, et cetera, and even arrive at q-calculus analogues for all of the usual functions one would want to have, such as an analogue for the sine function whose q-derivative is the appropriate analogue for the cosine.

History

The h-calculus is just the calculus of finite differences, which had been studied by George Boole and others, and has proven useful in a number of fields, among them combinatorics and fluid mechanics. The q-calculus, while dating in a sense back to Leonhard Euler and Carl Gustav Jacobi, is only recently beginning to see more usefulness in quantum mechanics, having an intimate connection with commutativity relations and Lie algebra.

Chapter 2

Calculation of Integrals

We know what integrals are the next problem is to find out how to do them. This turns out to be a heavy problem. For differentiation we had a tidy set of rules that allowed us to work out the derivative of just about any function that we cared to write down—the procedure is basically mechanical and can be done quite well by computers.

There is nothing like this for integration, in general. Even (apparently) simple integrals can be very difficult or downright impossible to work out.

More of this later. Integration is more of a skill than a routine. One approach that can work sometimes in simple cases is just 'spotting the answer'. If you can spot a function which differentiates to give your function then you have found an integral. Look at the following simple examples.

$$\frac{d}{dx}\left(\frac{1}{n+1}x^{n+1}\right) = x^n \quad \text{so} \quad \int x^n dx = \frac{1}{n+1}x^{n+1} \quad (n \neq -1)$$

$$\frac{d}{dx}\sin x = \cos x \quad \text{so} \quad \int \cos x\, dx = \sin x$$

$$\frac{d}{dx}\cos x = -\sin x \quad \text{so} \quad \int \sin x\, dx = -\cos x$$

$$\frac{d}{dx}e^{ax} = a^{ax} \quad \text{so} \quad \int e^{ax} dx = \frac{1}{a}e^{ax}$$

$$\frac{d}{dx}In\ x = \frac{1}{x} \quad so \quad \int \frac{dx}{x} = In\ x$$

Let me do something a little more complicated. We have seen that $\int\cos(x)\,dx = \sin(x)$. What then is the valu e of $\int\cos(2x)\,dx$?

We want to know what function differentiates to give $\cos(2x)$. Well, if you differentiate $\sin(2x)$ you get $2\cos(2x)$, which is not quite right because of the factor 2. That is easily adjusted for and we see that the required integral is $\sin(2x)/2$. Note that this simpe approach does not work with an integral like $\int\cos(x^2\,dx$.

$f(x)$	$\int f(x)dx$
$x^a a \neq 1$	$\frac{1}{a+1}x^a+1$
$\sin x$	$-\cos(x)$
$\cos(x)$	$\sin(x)$
e^x	e^x
$\frac{1}{x}$	$\text{In}(x)$
$\frac{1}{1+x^2}$	$\arctan x$
$\frac{1}{\sqrt{1-x^2}}$	$\arcsin x$

The obvious function to play with seems to be $\sin(x^2)$ but the derivative of this is $2x\cos(x^2)$ (damn the chain rule) which is nowhere near the required answer (and no, you can't go dividing by x!). Our simple approach has failed. Your work on differentiation will allow you to check that the following short table is correct.

These integrals should be known. There is a table of integrals at the end of this booklet.

There are books in the library that devote hndreds of pages to listing integrals. Most of the 'techniques' for working out integrals are just methods for changing one integral into anther in the hope that you will eventually come to an integral that you can 'spot' or find in the tables. Let me finish this section by raising (quietly) yet another difficulty with the indefinite integral—it really is plagued with problems. Consider the statement

$$\int \frac{dx}{x} = \ln(x)$$

which is in the above table. This looks correct because we have already found that the derivative of ln(x) is $1/x$. But look a little closer. The integrand ($1/x$) makes sense for any value of x other than 0. The integral (ln(x)) makes no sense at all if x 0. What we have in this formula is really an integral for $1/x$ valid for $x > 0$. We have had to limit the 'domain of definition' of the integrand in order to make sense. This problem will have to be coped with later on.

Properties of the Indefinite Integral

These are immediate consequences of the corresponding properties of derivatives. In each equation there is really an arbitrary constant of integration hanging around.

Let $f(x)$ and $g(x)$ be functions and λa constant. Then

$$\int \lambda f(x) dx = \lambda \int f(x) dx$$

$$\int f(x) + g(x) dx = \int f(x) dx + \int g(x) dx$$

provided that the integrals exist; we have not yet shown that there necessarily exists a function $F(x)$ that differentiates to give a specified function $f(x)$.

These rules may allow us to reduce an integral to the point where we can spot the answer.

$$\int 2x + e^x dx = 2\int x\,dx + \int e^x dx = x^2 + e^x$$

$$\int \sin x + 2\cos x\,dx = \int \sin x\,dx + 2$$

$$\int \cos x\,dx = -\cos x + 2\sin x$$

Impossible Integrals

You will frequently be told that some integrals are 'impossible' they cannot be done. Common examples are such seemingly respectable integrals as $\int e^{x2}$ dx, $\int \sin(x^2)$ dx,

$\int \frac{dx}{\sqrt{x^2+1}}$ *dx*. This does *not* mean that, for example, there is no function that differentiates to give e^{x2}; there certainly is. The problem is that you do not know how to write it down. The answer cannot be expressed in terms of 'elementary functions'.

Let me give another example to make this clearer. We have seen that

$$\int \frac{dx}{1+x^2} = \arctan(x)$$

That is simple enough. But suppose I had not done the inverse trig functions. Then you would never have met arctan(x), so I would have to say to you that, though the integral could certainly be done, it could not be done in terms of functions known to you. The same kind of thing applies to the 'impossible' examples given above. It is a distressing fact that the vast majority of integrals are, in this sense, impossible. When I write exercise sheets on differen-tiation I can get away with writing functions down at random, knowingfull well that they can be differentiated.

When writing exercises on integration I have to tread very carefully indeed and check everything out before I set anything. In some ways it is *more* surprising that derivatives *can* be done. It just happens to be the case that the derivatives of all the elementary functions can be expressed in terms of elementary functions. There is no logical reason for this—it is pure luck. That's a lie actually, there are good historical reasons why our idea of 'elementary' for functions leads to simple derivatives.

The Definite Integral

Suppose that $F(x)$ is an indefinite integral of $f(x)$, i.e. $F'(x) = f(x)$. The Definite Integral

$$\int_a^b f(x)\, dx$$

where a and b are numbers is defined to be the number $F(b) - F(a)$:

$$\int_a^b f(x)dx = \left[F(x)\right]_a^b = F(b) - F(a)$$

The choice of indefinite integral (choice of constant of integration does not matter—the constant of integration cancels out. a and b are called the Limits of Integration. a is called the Lower Limit and b is called the Upper Limit.

Note the 'square bracket' notation,

$$\left[f(x)\right]_a^b$$

stands for $f(b) - f(a)$. Some books use,

$$f(x)\Big|_a^b,$$

but this can be confusing in complicated expressions because it does not tell you where the expression starts. This may all seem a strange idea at this stage. The justification for definite integrals will come quite soon.

Differentiating an Integral

Consider the function,

$$g(x) = \int_a^x f(t)\,dt$$

where a is a constant and we are thinking of the upper limit of the integral as a variable. If $F(x)$ is an indefinite integral for $f(x)$ then

$$g(x) = F(x) - F(a)$$

So, by defiition of $F(x)$,

$$g'(x) = F'(x) = f(x)$$

This gives the result

$$\frac{d}{dx}\int_a^x f(t)\,dt = f(x)$$

Properties of the Definite Integral

If we assume that the function $f(x)$ has an integral then

$$\int_a^b f(x)\,dx = -\int_b^a f(x)\,dx \quad (F(b) - (F(a) = -(F(a) - (F(b)$$

$$\int_a^b f(x)\,dx + \int_a^c f(x)\,dx = -\int_a^c f(x)\,dx \quad ((F(b) - F(a))$$

$$+ (F(c) - (F(b)) = (F(c) - (F(a))$$

Infinite Limits

We have seen what we mean by a definite integral where both limits of integration are *finite*. In many important applications of integration you will come across cases where one or more of the limits of integration is infinite. There are three possibilities:

$$\int_a^{\infty} f(x)\, dx \quad \int_{-\infty}^{a} f(x)\, dx \quad \int_{-\infty}^{\infty} f(x)\, dx$$

The interpretation of these is that they are to be taken as limiting cases of the corresponding integrals with finite limits. For example,

$$\int_a^{\infty} f(x)\, dx = \lim_{b\to\infty} \int_a^b f(x)\, dx$$

$$\int_{-\infty}^{a} f(x\, dx = \lim_{b\to-\infty} \int_b^a f(x)\, dx$$

Here is a simple example: onsider the integral $\int_1^{\infty} \frac{dx}{x^2}$

Now, for any finite positive value f b,

$$\int_1^b \frac{dx}{x^2} = \left[\frac{-1}{x}\right]_1^b = 1 - \frac{1}{b}$$

Now we look at the limiting behaviour as $b\to\infty$. In the limit $1/b\to 0$ and so the value of the integral tends to 1. Therefore

$$\int_1^{\infty} \frac{dx}{x^2} = 1$$

Of course, in most cases the limits will not exist, so the integral will not exist either. For example,

$$\int_0^{\infty} x\, dx = \lim_{a\to\infty} \frac{1}{2} a^2$$

This limit does not exist, so the integral does not exist, though you might get away with saying that it is 'infinite'.

More seriously, consider the integral

$$\int_0^{\infty} \cos x\, dx = \lim_{a\to\infty} \sin a$$

Here the limit does not exist either, nor does it do anything so straightforward as going off to infinity—it just oscillates to and fro between –1 and 1.

Finally, let me note a small problem. The integral,

$$\int_{-\infty}^{\infty} f(x)$$

d*x* is the limiting case of,

$\int_a^b$ *f*(*x*) *dx* as *a* ® ¥ and *b* ® ¥.

This need not be the same thing as,

$$\lim_{a\to\infty} \int_{-a}^{a} f(x)dx.$$

It is best to think of such doubly-infinite limits in two stages: first let the top limit tend to infinity to get $\int_a^{\infty}$ and then let the bottom limit tend to - ¥ to get the answer.

The reason for such care is shown by th integral of sin(*x*). For any finite value of *a* we have,

$\int_{-a}^{a}$ in(*x*) d*x* = 0 (check).

If we took the limit *a* → ¥ in this we would get

$\int_{-\infty}^{\infty}$ sn(*x*) d*x* = 0

but this is actually nonsense. If, on the other hand, you start by studying $\int_a^{\infty}$ sin(*x*) d*x* you will see at once that no meaningful limiting value exists.

Chapter 3

Basics of Differential Calculus

We will be dropping the initial value of the index in quite a few facts and theorems that we'll be seeing throughout this chapter. In these facts/theorems the starting point of the series will not affect the result and so to simplify the notation and to avoid giving the impression that the starting point is important we will drop the index from the notation.

Do not forget however, that there is a starting point and that this will be an infinite series. Note however, that if we do put an initial value of the index on a series in a fact/theorem it is there because it really does need to be there. Now that some of the notational issues are out of the way we need to start thinking about various ways that we can manipulate series. We'll start this off with basic arithmetic with infinite series as we'll need to be able to do that on occasion. We have the following properties.

Properties

If $\sum a_n$ and $\sum b_n$ are both convergent series then,

1. $\sum ca_n$, where c is any number, is also convergent and

$$\sum ca_n = c\sum a_n$$

2. $\sum_{n=k}^{\infty} a_n \pm \sum_{n=k}^{\infty} b_n$ is also convergent and,

$$\sum_{n=k}^{\infty} a_n \pm \sum_{n=k}^{\infty} b_n = \sum_{n=k}^{\infty} (a_n \pm b_n).$$

The first property is simply telling us that we can always factor a multiplicative constant out of an infinite series and again recall that if we don't put in an initial value of the index that the series can start at any value. Also recall that in these cases we won't put an infinity at the top either.

The second property says that if we add/subtract series all we really need to do is add/subtract the series terms. Note as well that in order to add/subtract series we need to make sure that both have the same initial value of the index and the new series will also start at this value. Before we move on to a different topic let's discuss multiplication of series briefly. We'll start both series at $n = 0$ for a later formula and then note that,

$$\left(\sum_{n=0}^{\infty} a_n\right)\left(\sum_{n=0}^{\infty} b_n\right) \neq \sum_{n=0}^{\infty} (a_n b_n)$$

To convince yourself that this isn't true consider the following product of two finite sums.

$$(2 + x)(3 - 5x + x^2) = 6 - 7x - 3x^2 + x^3$$

Yeah, it was just the multiplication of two polynomials. Each is a finite sum and so it makes the point. In doing the multiplication we didn't just multiply the constant terms, then the x terms, *etc.* Instead we had to distribute the 2 through the second polynomial, then distribute the x through the second polynomial and finally combine like terms. Multiplying infinite series (even though we said we can't think of an infinite series as an infinite sum) needs to be done in the same manner. With multiplication we're really asking us to do the following,

$$\left(\sum_{n=0}^{\infty} a_n\right)\left(\sum_{n=0}^{\infty} b_n\right) = (b_0 + b_1 + b_2 + b_3 + \ldots)$$

To do this multiplication we would have to distribute the a_0 through the second term, distribute the a_1 through, etc then combine like terms. This is pretty much impossible since both series have an infinite set of terms in them, however the following formula can be used to determine the product of two series.

$$\left(\sum_{n=0}^{\infty} a_n\right)\left(\sum_{n=0}^{\infty} b_n\right) = \sum_{n=0}^{\infty} c_n \text{ where } cn = \sum_{n=0}^{\infty} a_i b_{n+1}$$

We also can't say a lot about the convergence of the product. Even if both of the original series are convergent it is possible for the product to be divergent.

The reality is that multiplication of series is a somewhat difficult process and in general is avoided if possible. We will take a brief look at it towards the end of the chapter when we've got more work under our belt and we run across a situation where it might actually be what we want to do. Until then, don't worry about multiplying series. The next topic that we need to discuss in this section is that of index shift. To be honest this is not a topic that we'll see all that often in this course. In fact, we'll use it once in the next section and then not use it again in all likelihood.

Despite the fact that we won't use it much in this course doesn't mean however that it isn't used often in other classes where you might run across series. So, we will cover it briefly here so that you can say you've seen it. The basic idea behind index shifts is to start a series at a different value for whatever the reason (and yes, there are legitimate reasons for doing that).

Consider the following series,

$$\sum_{i=2}^{\infty} \frac{n+5}{2^n}$$

Suppose that for some reason we wanted to start this series at $n = 0$, but we didn't want to change the value of the series. This means that we can't just change the $n = 2$ to $n = 0$ as this would add in two new terms to the series and thus changing its value. Performing an index shift is a fairly simple process to do.

We'll start by defining a new index, say i, as follows,

$$i = n - 2$$

Now, when $n = 2$, we will get $i = 0$. Notice as well that if $n = \infty$ then $i = \infty - 2$, so only the lower limit will change here. Next, we can solve this for n to get,

$$n = i + 2$$

We can now completely rewrite the series in terms of the index i instead of the index n simply be plugging in our equation for n in terms of i.

$$\sum_{i=2}^{\infty} \frac{n+5}{2^n} = \sum_{i=0}^{\infty} \frac{(i+2)+5}{2^{i+2}} = \sum_{i=0}^{\infty} \frac{i+7}{2^{i+2}}$$

To finish the problem out we'll recall that the letter we used for the index doesn't matter and so we'll change the final i back into an n to get,

$$\sum_{i=2}^{\infty} \frac{n+5}{2^n} = \sum_{i=0}^{\infty} \frac{n+7}{2^{n+2}}$$

To convince yourselves that these really are the same summation let's write out the first couple of terms for each of them,

$$\sum_{i=2}^{\infty} \frac{n+5}{2^n} = \frac{7}{2^2} + \frac{8}{2^3} + \frac{9}{2^4} + \frac{10}{2^5} + \ldots$$

$$\sum_{i=2}^{\infty} \frac{n+7}{2^{n+2}} = \frac{7}{2^2} + \frac{8}{2^3} + \frac{9}{2^4} + \frac{10}{2^5} + \ldots$$

So, sure enough the two series do have exactly the same terms.

There is actually an easier way to do an index shift. The method given above is the technically correct way of doing an index shift. However, notice in the above example we decreased the initial value of the index by 2 and all the n's in the series terms increased by 2 as well. This will always work in this manner. If we decrease the initial value of the index by a set amount then all the other n's in the series term will increase by the same amount. Likewise, if we increase the initial value of the index by a set amount, then all the n's in the series term will decrease by the same amount. Let's do a couple of examples using this shorthand method for doing index shifts.

Example: Perform the following index shifts.

1. Write $\sum_{i=1}^{\infty} ar^{n-1}$ as a series that starts at $n = 0$.

2. Write $\sum_{i=1}^{\infty} \frac{n^2}{1-3^{n+1}}$ as a series that starts at $n = 3$.

Solution:

1. In this case we need to decrease the initial value by 1 and so the n's (okay the single n) in the term must increase by 1 as well.

$$\sum_{i=1}^{\infty} ar^{n-1} = \sum_{i=1}^{\infty} ar^{(n+1)-1} = \sum_{i=1}^{\infty} ar^{n}$$

2. For this problem we want to increase the initial value by 2 and so all the n's in the series term must decrease by 2.

$$\sum_{i=1}^{\infty} \frac{n^2}{1-3^{n+1}} = \sum_{i=3}^{\infty} \frac{(n-2)^2}{1-3^{(n-2)+1}} = \sum_{i=3}^{\infty} \frac{(n-2)^2}{1-3^{n-1}}$$

The final topic in this section is again a topic that we'll not be seeing all that often in this class, although we will be seeing it more often than the index shifts.

This final topic is really more about alternate ways to write series when the situation requires it. Let's start with the following series and note that the $n = 1$ starting point is only for convenience since we need to start the series somewhere.

$$\sum_{n=1}^{\infty} a_n = a_1 + a_2 + a_3 + a_4 + a_5 + ...$$

Notice that if we ignore the first term the remaining terms will also be a series that will start at $n = 2$ instead of $n = 1$ So, we can rewrite the original series as follows,

$$\sum_{n=1}^{\infty} a_n = a_1 + \sum_{n=2}^{\infty} a_n$$

In this example we say that we've stripped out the first term. We could have stripped out more terms if we wanted to. In the following series we've stripped out the first two terms and the first four terms respectively.

$$\sum_{n=1}^{\infty} a_n = a_1 + a_2 + \sum_{n=3}^{\infty} a_n$$

$$\sum_{n=1}^{\infty} a_n = a_1 + a_2 + a_3 + a_4 + \sum_{n=5}^{\infty} a_n$$

Being able to strip out terms will, on occasion, simplify our work or allow us to reuse a prior result so it's an important idea to remember. Notice that in the second example above we could have also denoted the four terms that we stripped out as a finite series as follows,

$$\sum_{n=1}^{\infty} a_n = a_1 + a_2 + a_3 + a_4 + \sum_{n=5}^{\infty} a_n = \sum_{n=1}^{4} a_n + \sum_{n=5}^{\infty} a_n$$

This is a convenient notation when we are stripping out a large number of terms or if we need to strip out an undetermined number of terms. In general, we can write a series as follows,

$$\sum_{n=1}^{\infty} a_n = \sum_{n=1}^{N} a_n + \sum_{n=N+1}^{\infty} a_n$$

We'll leave this section with an important warning about terminology. Don't get sequences and series confused! A sequence is a list of numbers written in a specific order while an infinite series is a limit of a sequence of finite series and hence, if it exists will be a single value. So, once again, a sequence is a list of numbers while a series is a single number, provided it makes sense to even compute the series.

Students will often confuse the two and try to use facts pertaining to one on the other. However, since they are different beasts this just won't work. There will be problems where we are using both sequences and series so we'll always have to remember that they are different.

Convergence/Divergence

In the previous section we spent some time getting familiar with series and we briefly defined convergence and divergence. Before worrying about convergence and divergence of a series we wanted to make sure that we've started to get comfortable with the notation involved in series and some of the some of the various manipulations of series that we will, on occasion, need to be able to do. As noted in the previous section most of what we were doing there won't be done much in this chapter. So, it is now time to start talking about the convergence and divergence of a series as this will a topic that we'll be dealing with to one extent of another in almost all of the remaining sections of this chapter. So, let's recap just what an infinite series is and what it means for a series to be convergent or divergent.

We'll start with a sequence $\{a_n\}_{n=1}^{\infty}$ and again note that we're starting the sequence at $n = 1$ only for the sake of convenience and it can, in fact, be anything. Next we define the partial sums of the series as,

$$\begin{aligned} s_1 &= a_1 \\ s_2 &= a_1 + a_2 \\ s_3 &= a_1 + a_2 + a_3 \\ s_4 &= a_1 + a_2 + a_3 + a_4 \\ &\vdots \\ s_n &= a_1 + a_2 + a_3 + a_4 + \dots + a_n = \sum_{i=1}^{n} a_i \end{aligned}$$

and these form a new sequence, $\{s_n\}_{n=1}^{\infty}$.

An infinite series, or just series here since almost every series that will be looking at will be an infinite series, is then the limit of the partial sums. Or,

$$\sum_{i=1}^{\infty} a_i = \lim_{n\to\infty} s_n$$

If the sequence of partial sums is a convergent sequence (*i.e.* its limit exists and is finite) then the series is also called convergent and in this case if $\lim_{n\to\infty} s_n = s$ then, $\sum_{i=1}^{\infty} a_i = s$. Likewise, if the sequence of partial sums is a divergent sequence (i.e. its limit doesn't exist or is plus or minus infinity) then the series is also called divergent. Let's take a look at some series and see if we can determine if they are convergent or divergent and see if we can determine the value of any convergent series we find.

Example: Determine if the following series is convergent or divergent. If it converges determine its value.

$$\sum_{i=1}^{\infty} n$$

Solution: To determine if the series is convergent we first need to get our hands on a formula for the general term in the sequence of partial sums.

$$s_n = \sum_{i=1}^{n} i$$

This is a known series and its value can be shown to be,

$$sn = \sum_{i=1}^{n} i = \frac{n(n+1)}{2}$$

Don't worry if you didn't know this formula (I'd be surprised if anyone knew it...) as you won't be required to know it in my course. So, to determine if the series is convergent we will first need to see if the sequence of partial sums,

$$\left\{\frac{n(n+1)}{2}\right\}_{n=4}^{\infty}$$

Is convergent or divergent. That's not terribly difficult in this case. The limit of the sequence terms is,

$$\lim_{n\to\infty} \frac{n(n+1)}{2} = \infty$$

Therefore, the sequence of partial sums diverges to and so the series also diverges.

So, as we saw in this example we had to know a fairly obscure formula in order to determine the convergence of this series. In general finding a formula for the general term in the sequence of partial sums is a very difficult process.

In fact after the next section we'll not be doing much with the partial sums of series due to the extreme difficulty faced in finding the general formula.

This also means that we'll not be doing much work with the value of series since in order to get the value we'll also need to know the general formula for the partial sums. We will continue with a few more examples however, since this is technically how we determine convergence and the value of a series. Also, the remaining examples we'll be looking at in this section will lead us to a very important fact about the convergence of series. So, let's take a look at a couple more examples.

Example: Determine if the following series converges or diverges. If it converges determine its sum.

$$\sum_{n=2}^{\infty} \frac{1}{n^2 - 1}$$

Solution: This is actually one of the few series in which we are able to determine a formula for the general term in the sequence of partial fractions. However, in this section we are more interested in the general idea of convergence and divergence and so we'll put off discussing the process for finding the formula until the next section. The general formula for the partial sums is,

$$s_n = \sum_{i=2}^{n} \frac{1}{i^2 - 1} = \frac{3}{4} - \frac{1}{2n} - \frac{1}{2(n+1)}$$

and in this case we have,

$$\lim_{n\to\infty} s_n = \lim_{n\to\infty}\left(\frac{3}{4}-\frac{1}{2n}-\frac{1}{2(n+1)}\right)=\frac{3}{4}$$

The sequence of partial sums converges and so the series converges also and its value is,

$$\sum_{i=2}^{n}\frac{1}{n^2-1} = \frac{3}{4}$$

Example: Determine if the following series converges or diverges. If it converges determine its sum.

$$\sum_{n=0}^{\infty}(-1)^n$$

Solution: In this case we really don't need a general formula for the partial sums to determine the convergence of this series. Let's just write down the first few partial sums.

$$\begin{aligned} s_0 &= 1 \\ s_1 &= 1-1=0 \\ s_2 &= 1-1+1=1 \\ s_3 &= 1-1+1-1=0 \ \ etc. \end{aligned}$$

So, it looks like the sequence of partial sums is,

$$\{s_n\}_{n=0}^{\infty} = \{1, 0, 1, 0, 1, 0, 1, 0, 1,\ldots\}$$

and this sequence diverges since $\lim\limits_{n\to\infty} s_n$ doesn't exist. Therefore, the series also diverges.

Example: Determine if the following series converges or diverges. If it converges determine its sum.

$$\sum_{n=1}^{\infty}\frac{1}{3^{n-1}}$$

Solution: Here is the general formula for the partial sums for this series.

$$sn = \sum_{i=1}^{n} \frac{1}{3^{i-1}} = \frac{3}{2}\left(1 - \frac{1}{3^n}\right)$$

Again, do not worry about knowing this formula. This is not something that you'll ever be asked to know in my class.

In this case the limit of the sequence of partial sums is,

$$\lim_{n\to\infty} s_n = \lim_{n\to\infty} \frac{3}{2}\left(1 - \frac{1}{3^n}\right) = \frac{3}{2}$$

The sequence of partial sums is convergent and so the series will also be convergent. The value of the series is,

$$\sum_{i=1}^{\infty} \frac{1}{3^{n-1}} = \frac{3}{2}$$

As we already noted, do not get excited about determining the general formula for the sequence of partial sums. There is only going to be one type of series where you will need to determine this formula and the process in that case isn't too bad. In fact, you already know how to do most of the work in the process as you'll see in the next section.

So, we've determined the convergence of four series now. Two of the series converged and two diverged. Let's go back and examine the series terms for each of these. For each of the series let's take the limit as n goes to infinity of the series terms (not the partial sums!!).

$\lim_{n\to\infty} n = \infty$ this series diverged

$\lim_{n\to\infty} \frac{1}{n^2 - 1} = 0$ this series converged

$\lim_{n\to\infty} (-1)^n$ doesn t exist this series diverged

$\lim_{n\to\infty} \frac{1}{3^{2-1}} = 0$ this series converged

Notice that for the two series that converged the series term itself was zero in the limit. This will always be true for convergent series and leads to the following theorem.

Theorem: If $\sum a_n$ converges then $\lim_{n\to\infty} a_n = 0$.

Proof: First let's suppose that the series starts at $n = 1$. If it doesn't then we can modify things as appropriate below. Then the partial sums are,

$$s_{n-1} = \sum_{i=1}^{n-1} a_i = a_1 + a_2 + a_3 + a_4 + \ldots + a_{n-1}$$

$$s_n = \sum_{i=1}^{n} a_i = a_1 + a_2 + a_3 + a_4 + \ldots + a_{n-1} + a_n$$

Next, we can use these two partial sums to write,

$$a_n = s_n - s_{n-4}$$

Now because we know that $\sum a_n$ is convergent we also know that the sequence $\{s_n\}_{n=1}^{\infty}$ is also convergent and that $\lim_{n\to\infty} s_n = s$ for some finite value s. However, since $n-1 \to \infty$ as $n \to \infty$ $n \to \infty$ we also have $\lim_{n\to\infty} s_{n-1} = s$.

We now have,

$$\lim_{n\to\infty} a_n = \lim_{n\to\infty} (s_n - s_{n-1}) = \lim_{n\to\infty} s_n - \lim_{n\to\infty} s_{n-1} = s - s = 0$$

Be careful to not misuse this theorem! This theorem gives us a requirement for convergence but not a guarantee of convergence. In other words, the converse is NOT true. If $\lim_{n\to\infty} a_n = 0$ the series may actually diverge! Consider the following two series.

$$\sum_{n=1}^{\infty} \frac{1}{n} \qquad \sum_{n=1}^{\infty} \frac{1}{n^2}$$

In both cases the series terms are zero in the limit as n goes to infinity, yet only the second series converges. The first series diverges. It will be a couple of sections before we can prove this, so at this point please believe this and know that you'll be able to prove the convergence of these two series in a couple of sections.

Again, as noted above, all this theorem does is give us a requirement for a series to converge. In order for a series to converge the series terms must go to zero in the limit. If the series terms do not go to zero in the limit then there is no way the series can converge since this would violate the theorem.

Divergence Test

If $\lim_{n \to \infty} a_n \neq 0$ then $\sum a_n$ will diverge.

Again, do NOT misuse this test. This test only says that a series is guaranteed to diverge if the series terms don't go to zero in the limit. If the series terms do happen to go to zero the series may or may not converge! Again, recall the following two series,

$$\sum_{n=1}^{\infty} \frac{1}{n} \quad \text{diverges}$$

$$\sum_{n=1}^{\infty} \frac{1}{n^2} \quad \text{converges}$$

One of the more common mistakes that students make when the first get into series is to assume that if $\lim_{n \to \infty} a_n = 0$ $\lim_{n \to \infty} a_n = 0$ then $\sum a_n$ will converge. There is just no way to guarantee this so be careful!

Example: Determine if the following series is convergent or divergent.

$$\sum_{n=0}^{\infty} \frac{4n^2 \quad n^3}{10 + 2n^3}$$

Solution: With almost every series we'll be looking at in this chapter the first thing that we should do is take a look at the series terms and see if they go to zero of not. If it's clear that the terms don't go to zero use the Divergence Test and be done with the problem. That's what we'll do here.

$$\lim_{n \to \infty} \frac{4n^2 - n^3}{10 + 2n^3} = -\frac{1}{2} \neq 0$$

The limit of the series terms isn't zero and so by the Divergence Test the series diverges. The divergence test is the first test of many tests that we will be looking at over the course of the next several sections. You will need to keep track of all these tests, the conditions under which they can be used and their conclusions all in one place so you can quickly refer back to them as you need to. Next we should talk briefly revisit arithmetic of series and convergence/divergence. As we saw in the previous section if $\sum a_n$ and $\sum b_n$ are both convergent series then so are $\sum ca_n$ and $\sum_{n=k}^{\infty}(a_n \pm b_n)$. Furthermore, these series will have the following sums or values.

$$\sum ca_n = c\sum a_n$$

$$\sum_{n=k}^{\infty}(a_n \pm b_n) = \sum_{n=k}^{\infty} a_n \pm \sum_{n=k}^{\infty} b_n$$

We'll see an example of this in the next section after we get a few more examples under our belt.

At this point just remember that a sum of convergent sequences is convergent and multiplying a convergent sequence by a number will not change its convergence. We need to be a little careful with these facts when it comes to divergent series.

In the first case if $\sum a_n$ is divergent then $\sum ca_n$ will also be divergent (provided c isn't zero of course) since multiplying a series that is infinite in value or doesn't have a value by a finite value (*i.e.* c) won't change the fact that the series has an infinite or no value. However, it is possible to have both $\sum a_n$ and $\sum b_n$ be divergent series and yet have,

$$\sum_{n=k}^{\infty}(a_n \pm b_n)$$

be a convergent series. Now, since the main topic of this section is the convergence of a series we should mention a stronger type of convergence.

A series $\sum a_n$ is said to converge absolutely if $\sum |a_n|$ also converges. Absolute convergence is stronger than convergence in the sense that a series that is absolutely convergent will also be convergent, but a series that is convergent may or may not be absolutely convergent.

In fact if $\sum a_n$ converges and $\sum |a_n|$ diverges the series $\sum a_n$ is called conditionally convergent.

At this point we don't really have the tools at hand to properly investigate this topic in detail nor do we have the tools in hand to determine if a series is absolutely convergent or not. So we'll not say anything more about this subject for a while. When we finally have the tools in hand to discuss this topic in more detail we will revisit it. Until then don't worry about it. The idea is mentioned here only because we were already discussing convergence in this section and it ties into the last topic that we want to discuss in this section.

In the previous section after we'd introduced the idea of an infinite series we commented on the fact that we shouldn't think of an infinite series as an infinite sum despite the fact that the notation we use for infinite series seems to imply that it is an infinite sum. It's now time to briefly discuss this. First, we need to introduce the idea of a rearrangement.

A rearrangement of a series is exactly what it might sound like, it is the same series with the terms rearranged into a different order.

For example, consider the following the infinite series.

$$\sum_{n=1}^{\infty} a_n \quad = a_1 + a_2 + a_3 + a_4 + a_5 + a_6 + a_7 + \dots$$

A rearrangement of this series is,

$$\sum_{n=1}^{\infty} a_n \quad = a_2 + a_1 + a_3 + a_{14} + a_5 + a_9 + a_4 + \dots$$

The issue we need to discuss here is that for some series each of these arrangements of terms can have a different values despite the fact that they are using exactly the same terms.

Here is an example of this. It can be shown that,

$$\sum_{n=1}^{\infty}\frac{(-1)^{n+1}}{n} = 1-\frac{1}{2}+\frac{1}{3}-\frac{1}{4}+\frac{1}{5}-\frac{1}{6}+\frac{1}{7}-\frac{1}{8}+\ldots = \text{In}\,2$$

Since this series converges we know that if we multiply it by a constant c its value will also be multiplied by c. So, let's multiply this by,

$$\frac{1}{2}$$

to get,

$$\frac{1}{2}-\frac{1}{4}+\frac{1}{6}-\frac{1}{8}+\frac{1}{10}-\frac{1}{12}+\frac{1}{14}+\frac{1}{16}\ldots=\frac{1}{2} \quad = \text{In } 2$$

Now, let's add in a zero between each term as follows.

$$0+\frac{1}{2}+0-\frac{1}{4}+0+\frac{1}{6}+0-\frac{1}{8}+0+\frac{1}{10}+0\frac{1}{12}+0+\ldots=\frac{1}{2}\ \text{In } 2$$

We know that if two series converge we can add them by adding term by term

$$1+\frac{1}{3}-\frac{1}{2}+\frac{1}{5}+\frac{1}{7}-\frac{1}{4}+\ldots=\frac{3}{2}\ \text{In } 2$$

Facts

Given the series $\sum a_n$,

1. If $\sum a_n$ is absolutely convergent and its value is s then any rearrangement of $\sum a_n$ will also have a value of s.
2. If $\sum a_n$ is conditionally convergent and r is any real number then there is a rearrangement of $\sum a_n$ whose value will be r.

Again, we do not have the tools in hand yet to determine if a series is absolutely convergent and so don't worry about this at this point. This is here just to make sure that you understand that we have to be very careful in thinking of an infinite series as an infinite sum.

There are times when we can (i.e. the series is absolutely convergent) and there are times when we can't (i.e. the series is conditionally convergent).

As a final note, the fact above tells us that the series,

$$\sum_{n=1}^{\infty} \frac{(-1)^{n+1}}{n}$$

must be conditionally convergent since two rearrangements gave two separate values of this series. Eventually it will be very simple to show that this series conditionally convergent.

Special Series

In this section we are going to take a brief look at three special series. Actually, special may not be the correct term. All three have been named which makes them special in some way, however the main reason that we're going to look at two of them in this section is that they are the only types of series that we'll be looking at for which we will be able to get actual values for the series.

The third type is divergent and so won't have a value to worry about.

In general, determining the value of a series is very difficult and outside of these two kinds of series that we'll look at in this section we will not be determining the value of series in this chapter. So, let's get started.

Chapter 4

Vector Calculus

Linear relationships are those that produce a straight line when plotted on a graph. They are used to describe an enormous number of situations, from the slope of a road to the expansion of the universe. However straight lines are not enough to describe fully the complicated world in which we live.

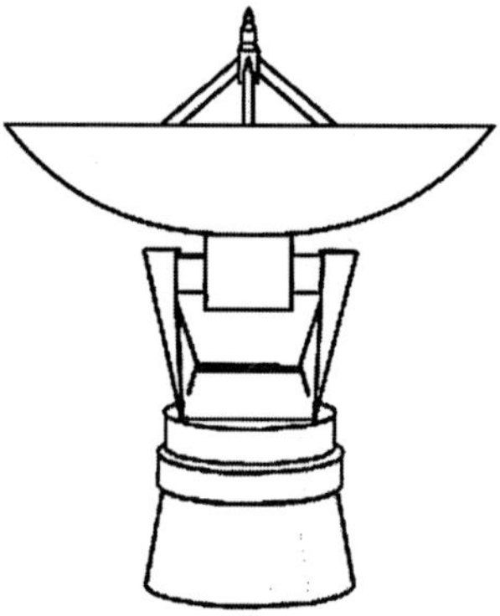

Figure: *The Satellite Dish is a Curve.*

For example, the smooth shape of the satellite dish is a curve described by quadratic equation. In this unit you'll explore the properties of quadratic equations and learn how to plot them.

Ellipse

For the points $P(x, y)$ on an ellipse the sum of the distances from two fixed points $F_{\pm}$ (the foci) is a given positive constant:

$$|\overrightarrow{PF_-}| + |\overrightarrow{PF_+}| = 2a$$

with $2a > |\overrightarrow{F_-F_+}|$.

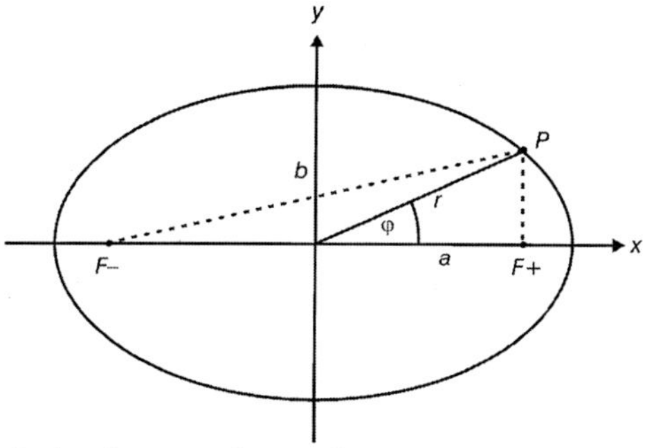

If $F_\pm = (\pm f, 0)$, then we have for the coordinates

$$\frac{x^2}{a^2} + \frac{y^2}{b^2} = 1,\ b^2 = a^2 - f^2,$$

and

$$r^2 = \frac{b^2}{1-(f/a)^2\cos^2(\varphi)}$$

for the polar coordinates of the points P.

A parameterisation of the ellipse is given by

$x = a \cos t, \quad y = b \sin t$

with $t \in [0, 2\pi)$.

Equivalence of the representations follows immediatly by calculation.

To proof

$$|\overrightarrow{PF_-}| + |\overrightarrow{PF_+}| = 2a \quad \Leftrightarrow \quad \frac{x^2}{a^2} + \frac{y^2}{b^2} = 1, \quad b^2 = a^2 - f^2$$

by squaring of,

$$\underbrace{2a - \sqrt{(x+f)^2 + y^2}}_{>0} = \sqrt{(x-f)^2 + y^2}$$

we obtain the equivalent equation

$$\underbrace{4a^2 + 4xf}_{>0} = 4a\sqrt{(x-f)^2 + y^2}\,.$$

Squaring again and dividing by $4a$ yield

$$a^2 + 2xf + \frac{f^2}{a^2}x^2 = x^2 + 2xf + f^2 + y^2.$$

Substituting $f^2 = a^2 - b^2$ and transforming we obtain the coordinate form.

To proof

$$r^2 = \frac{b^2}{1-(f/a)^2\cos^2(\varphi)}$$

we multiply by the denominator and take into account that

$x^2 = (x^2 + y^2)\cos^2(\varphi)$

This implies

$$x^2 + y^2 - \frac{a^2 - b^2}{a^2}x^2 = b^2$$

and division by b^2 yields the coordinate form.

Focus Rays

A ray of light originating at a focus will pass through the other focus after a simple reflection.

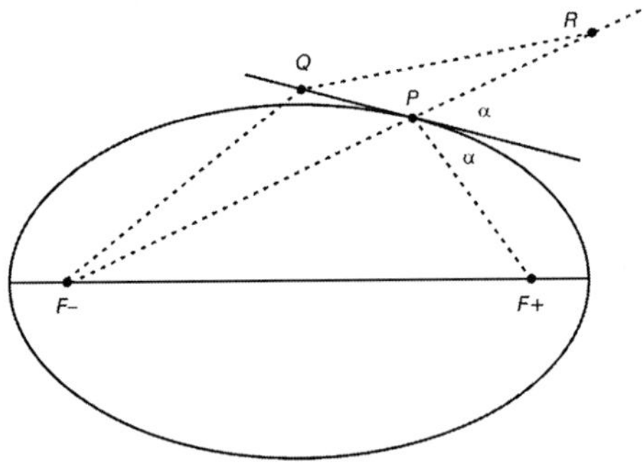

To prove this, select the following auxiliary points: the mirror image R of focus F_+ with respect to tangent and an arbitrary $Q \neq P$ point on g. Since Q lies outside of the ellipse we have

$$2a = |\overrightarrow{PF_-}| + |\overrightarrow{PF_+}| < |\overrightarrow{QF_-}| + |\overrightarrow{PF_+}|.$$

Sustituting the line segments $\overrightarrow{PF_+}$ and $\overrightarrow{QF_+}$ by its mirror images we obtain

$$|\overrightarrow{PF\_}|+|\overrightarrow{PR}|<|\overrightarrow{QF\_}|+|\overrightarrow{QR}|,$$

and hence the points $F\_$, P, R lie on a line. Thus we have $\sphericalangle(F\_PQ)=\alpha$.

This reflection property is used to break up kidney stones (lithotripsy). The rays of a radial source located in focus $F\_$ can be focused in F_+ by an elliptic reflector (spheroid).

Parabola

The points $P=(x, y)$ of a parabola have the same distance from a given point (focus) and a given line g (directrix).

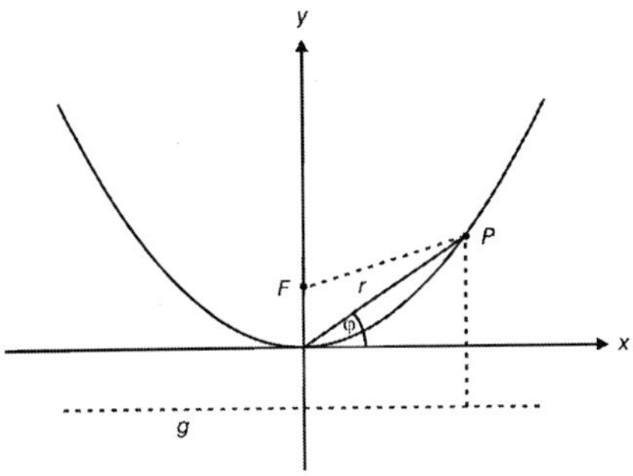

If $F=(0, f)$ and $g: y=-f$, then we have for the coordinates $4fy=x^2$

and

$$r=\frac{4f\sin(\varphi)}{\cos^2(\varphi)}$$

for the polar coordinates of the points P. Equivalence of the representations is evident. Equating the squared distances,

$$|\overrightarrow{PF}|^2 = x^2+(y-f)^2=(y+f)^2 \text{ dist }(P, g),$$

we obtain the coordinate form. The substitution yields the polar form.

Satellite TV

Parallel rays perpendicular to the directrix are bundled in the focus. This property is used to amplify incoming signals in parabol aerials ('satellite dishes').

To prove this property, first of all, we see that

$$\overrightarrow{FP} + \overrightarrow{QP} = \left(\begin{pmatrix} x \\ y \end{pmatrix} - \begin{pmatrix} 0 \\ f \end{pmatrix}\right) + \left(\begin{pmatrix} x \\ y \end{pmatrix} - \begin{pmatrix} x \\ -f \end{pmatrix}\right) = \begin{pmatrix} x \\ 2y \end{pmatrix}$$

is parallel to thedirection of the tangent in point P,

$$\begin{pmatrix} 1 \\ x/(2f) \end{pmatrix}.$$

Since $\qquad |\overrightarrow{FP}| = |\overrightarrow{FQ}|,$

it follows equality of the angles $\sphericalangle(F, P, R)$ and $\sphericalangle(R, P, Q)$, where R is the intersection point of the tangent and g.

Hyperbola

For the points $P = (x, y)$ on a hyperbola the difference of the distances from two given points (foci) $F_{\pm}$ is constant:

$$|\overrightarrow{PF_-}| - |\overrightarrow{PF_+}| = \pm 2a$$

with $\qquad 2a < |\overrightarrow{F_-F_+}|.$

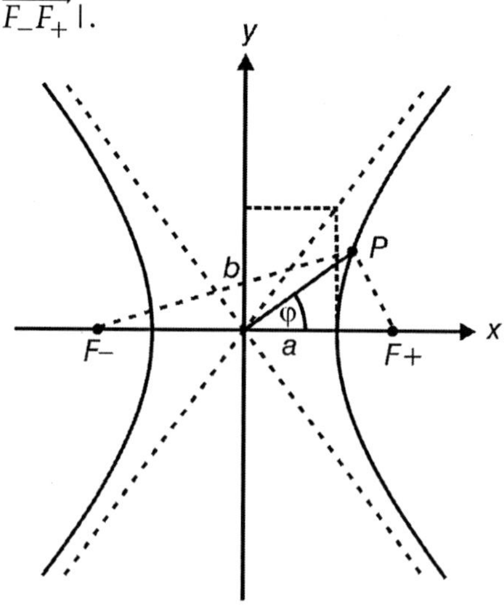

If $F_{\pm} = (\pm f, 0)$, then we have for the coordinates

$$\frac{x^2}{a^2} - \frac{y^2}{b^2} = 1, \quad b^2 = f^2 - a^2,$$

and $$r^2 = -\frac{b^2}{1-(f/a)^2 \cos^2 \theta}$$

for the polar coordinates of the points P. The asymptotes have the slope $\pm\, b/a$.

A parametrisation of the hyperbola branches is given by

$$x = \pm a \cosh t, \quad y = b \sinh t$$

with $r \in \mathbb{R}$.

Equivalence of the representations follows immediatly by calculation.

To proof

$$|\overrightarrow{PF_-}| - |\overrightarrow{PF_+}| = \pm\, 2a \Leftrightarrow \frac{x^2}{a^2} - \frac{y^2}{b^2} = 1, \; b^2 = a^2 - f^2$$

squaring of

$$\underbrace{\sqrt{(x+f)^2 + y^2} - 2a}_{>0} = \sqrt{(x-f)^2 + y^2}$$

leads to the equivalent equation

$$\underbrace{4a^2 + 4xf}_{>0} = 4a\sqrt{(x+f)^2 + y^2}\,.$$

Squaring again and dividing by $4a$ yield

$$a^2 + 2xf + \frac{f^2}{a^2}x^2 = x^2 + 2xf + f^2 + y^2.$$

By Substituting $f^2 = a^2 + b^2$ and transforming we obtain the coordinate form.

To proof

$$r^2 = -\frac{b^2}{1-(f/a)^2 \cos^2(\varphi)}$$

we multiply the polar form by the denominator and observe that

$$x^2 \ (x^2 + y^2) \cos^2 (\varphi).$$

Thus follows

$$x^2 + y^2 - \frac{a^2 - b^2}{a^2} x^2 = -b^2,$$

and division by $-b^2$ yields the coordinate form.

Navigation

To determine the position P of a ship, the time differences $2a_{j,k}$, while receiving synchronous radio signals of different transmitting stations F_i, are compared. Hence, the point P can be determined as point of intersection of hyperbolas with foci F_j, F_k.

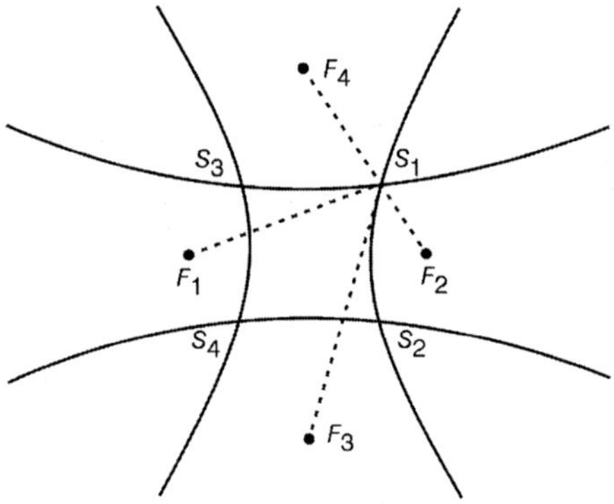

As a concrete example put

$$F_1 = (-2, 0), \quad F_2 = (0, 2), \quad F_3 = (0, -3), \quad F_4 = (3, 0)$$

and. $a_{1,2} = a_{3,4} = 1$ The corresponding equations of the hyperbolas are

$$x^2 - \frac{y^2}{3} = 1$$

$$x^2 - \frac{y^2}{8} = 1,$$

and we obtain $x_2 = \frac{32}{23}, y^2 = \frac{27}{23}$

for the coordinates of the possible points of intersection S_i.

Rotation of Conic Sections

If the conic section with the equation

$$ax^2 + bxy + cy^2 + dx + ey + f = 0$$

is transformed by a rotation about the origin through the angle α, then the equation in the new coordinates $\tilde{x}$ and $\tilde{y}$ has the form

$$\tilde{a}\tilde{x}^2 + \tilde{c}\tilde{y}^2 + \tilde{d}\tilde{x} + \tilde{e}\tilde{y} + \tilde{f} = 0,$$

provided

$$\tan 2\alpha = \frac{b}{a-c} \quad (a \neq c)$$

or $$\alpha = \frac{\pi}{4} \quad (a = c)$$

The coordinate transformation which expresses the old coordinates in terms of the new ones is given by

$$x = \tilde{x}\cos\alpha - \tilde{y}\sin\alpha, \qquad y = \tilde{x}\sin\alpha + \tilde{y}\cos\alpha.$$

The parameters of the new equation are

$$\tilde{a} = a\cos^2\alpha + b\cos\alpha\sin\alpha + c\sin^2\alpha,\ \tilde{d} = d\cos\alpha + e\sin\alpha,\ \tilde{b} = 0,$$

$$\tilde{c} = a\sin^2\alpha - b\cos\alpha\sin\alpha + c\cos^2\alpha,\ \tilde{e} = e\cos\alpha - d\sin\alpha,\ \tilde{f} = f.$$

Note that in the new equation there is no mixed quadratic term, i.e. the coeffcient $\tilde{b}$ of $\tilde{x}\tilde{y}$ is zero.

Thus this equation may be easily further transformed into a normal form of the conic section by completing squares.

Consider the conic section

$$x^2 - 2\sqrt{3}xy + 3y^2 - \sqrt{3x} - y + f = 0.$$

The condition on the angle α of a suitable rotation of the coordinate axes is

$$\tan 2a = \sqrt{3}.$$

Thus with $a = \frac{\pi}{6}$

the coordinate transformation

$$x = \sqrt{3}\tilde{x}/2 - \tilde{y}/2, \qquad y = \tilde{x}/2 + \sqrt{3}\tilde{y}/2$$

yields

$$\tilde{a} = 1\cdot 3/4 + (-2\sqrt{3})\cdot\sqrt{3}/4 + 3\cdot 1/4 = 0,$$

$$\tilde{c} = 1\cdot 1/4 - (-2\sqrt{3})\cdot\sqrt{3}/4 + 3\cdot 3/4 = 4,$$

$$\tilde{d} = -\sqrt{3}\cdot\sqrt{3}/2 + (-1)\cdot 1/2 = -2,$$

$$\tilde{e} = -1\cdot\sqrt{3}/2 - (-\sqrt{3})\cdot 1/2 = 0,$$

and the equation gets the form

$$4\tilde{y}^2 - 2\tilde{x} + f = 0.$$

Consequently the given conic section is a parabola (independent from f).

Curves and Surfaces

Every graphics system has some form of primitive to draw lines. Using these primitives we can draw many complex shapes. However, as these shapes get ever more complex and finely detailed, so does the data needed to describe them accurately. The worst case scenario is the curve. A curve can be described by a finite number of short straight segments. However, on close inspection this is only an approxi-mation. To get a better approximation we can use more segments per unit length. This increases the amount of data required to store the curve and makes it difficult to manipulate. We clearly need a way of representing these curves in a more mathematical fashion.

The types of curve will be discussed fall into two broad categories: interpolating or approximation curves. Interpolating curves will pass through the points used to describe it, whereas an approximating curve will get near to the points, though exactly what is called near will be discussed later.

The points through which the curve passes are known as knots; the curve described by the equation is often referred to as a spline. This term originated in manual design, where a spline is a thin strip. This strip was held in place by weights to create a curve which could then be traced. In the same way we now use knots to describe a curve.

Curves

A curve maybe represented parametrically by

$x = \psi(t),$

where x is a point on the curve, and t is a parameter that varies along the curve.

Remarks:

1. Pure mathematicians define a 'curve' (or 'path' if the curve is 'smooth' enough) to be a mapping, while the image of the mapping is a 'set of points of the curve'. Applied mathematicians tend to identify the set of points with the curve.
2. The parametrisation of a given curve is not unique.
3. We will concentrate on $\mathbb{R}^3$, but there are straightforward extensions to $\mathbb{R}^m$.
4. One of our aims is to identify what characterises a curve.

Examples:

1. $x = \psi(t) = (at, bt^2, 0)$.

   This is the parabola $a^2y = bx^2$ in the xy-plane.

2. $x = \psi(t) = (e^t \cos t, e^t \sin t, 0)$.

   This is a spiral.

3. $x = \psi(t) = (\cos t, \sin t,)$.

   This is a helix on the cylinder of radius one, i.e. the cylinder $\rho^2 = x^2 + y^2 = 1$.

4. Suppose $x = (\rho(t) \cos \phi(t), \rho(t) \sin \phi(t), z(t))$, where

$\rho = \gamma + \cos at, \phi = \beta t, z = \sin at,$

and α, β and $\gamma > 1$ are constants. This curve lies on the torus $(\rho - \gamma)^2 + z^2 = 1$.

Unlectured paragraph discussing when the curve is closed. For the curve defined by $0 \leq t \leq t_c$, to be closed it is necessary that the curve

1. Returns to $x = \gamma + 1$ at t_c, which implies that $(\gamma + \cos \alpha t_c) \cos \beta t_c = \gamma + 1$,
   i.e. that $\alpha t_c = p\pi$ and $\beta t_c = q\pi$, where p and q are integers;
2. Returns to $y = 0$ at t_c, which is ensured if $\beta t_c = q\pi$ since $\sin q\pi = 0$;
3. Returns to $z = 0$ at t_c, which is ensured if $\alpha t_c = \pi$ since $\sin p\pi = 0$.

It follows from the above three results that for the curve to be closed it is necessary that

$$\frac{\alpha}{\beta} = \frac{p}{q},$$

i.e. that α / β is rational.

Reading Exercise. What happens if α / β is irrational?

Definitions:

1. A simple curve is one that does not intersect or touch itself. simple intersects touches
2. A curve is smooth if the derivative $\psi'(t)$ exists. It is piecewise smooth if the curve can be divided into a finite number of smooth curves.

The Length of a Simple Curve

Let C be a simple curve defined by

$x = \psi(t)$ for $t_a \leq t \leq t_b$.

Form a dissection D by splitting $[t_a, t_b]$ into N intervals $t_a = t_0 < t_1 < \ldots < t_N = t_b$.

The points $\{\psi(t_0), \psi(t_1), \ldots \psi(t_N)\}$ can be viewed as the vertices of an inscribed polygonal path. The length of this path is

$$\ell_D = \sum_{j=1}^{N} |\psi(t_j)\,\psi(t_j - 1)|,.$$

Define $$\ell(C) = \sup_D \ell_D.$$

Then if $\ell(C) < \infty$, C is said to be rectifiable and $\ell(C)$ is the length of the curve.

Remarks:

1. If C is not simple, but can be split into a finite number of simple curves C_j of length $1(C_j)$, then

$$\ell(C) = \sum_j \ell(C_j).$$

2. If is continuously dierentiable (i.e. if the vector derivative $\psi'(t) = (\psi_1'(t), \psi_2'(t), \psi_3'(t))$ is continuous), then

$$\ell = \int_{t_a}^{t_b} |\psi'(t)|\, dt.$$

This result can be derived using the definition of a Riemann integral and the idea of uniform continuity.

The Arc Length

Definition: The arc length [function] of a smooth curve C parameterised by $\psi(t)$ is defined to be

$$s(t) = \int_{t_a}^{t_b} |\psi'(\tau)|\, d\tau.$$

Remarks:

1. At one end of the curve the arc length is zero, while at the other end the arc length equals the length of the curve,

$$s(t_a) = 0, \text{ and } s(t_b) = \ell.$$

2. The arc length is an increasing function since

$$\frac{ds}{dt} = |\psi'(t)| \geq 0.$$

3. Using $x = \psi(t)$, the infinitesimal arc length can be re-expressed (admittedly slightly cavalierly) as

$$\frac{ds}{dt} = |\psi'(t)|\, dt = \left[\left(\frac{dx}{dt}\right)^2 + \left(\frac{dy}{dt}\right)^2 + \left(\frac{dz}{dt}\right)^2\right]^{1/2}$$

$$dt = (dx^2 + dy^2 + dz^2)^{\frac{1}{2}}.$$

This is as expected on geometrical grounds. Further, on geometrical grounds we would expect that

$$\ell = \int (\mathrm{d}x^2 + dy^2 + dz^2)^{1/2}.$$

$$= \int_{t_a}^{t_b} |\psi'(t)|\, dt.$$

4. If $\frac{ds}{dt} = |\psi'(t)| > 0$ then (by the inverse function theorem this term's Analysis course) s(t) is invertible, and there is an inverse function such that

$$t = \lambda(s).$$

We can then parametrise the curve in terms of s:

$x = \psi\,(t) = \psi\,(l(s)) = r(s).$

Further, $\frac{d\lambda}{ds} = \frac{dt}{ds} = \frac{1}{ds/dt} = \frac{1}{|\psi'(t)|} = \frac{1}{|\psi'(\lambda(s))|}$, it follows from using the chain rule that

$$r'(s) = \psi'(\lambda(s))\frac{d\lambda}{ds} = \frac{\psi'(\lambda(s))}{|\psi'(\lambda(s))|},$$

and hence that (as required for consistency)

$| \, r'\,(s) \, | = 1$

Divergence Theorem

Line Integrals

In single-variable calculus you learned how to integrate a real-valued function $f\,(x)$ over an interval $[a, b]$ in ¡^{1}.

This integral (usually called a *Riemann integral*) can be thought of as an integral over a *path* in ¡^{1}, since an interval (or collection of intervals) is really the only kind of "path" in ¡^{1}.

You may also recall that if $f\,(x)$ represented the force applied along the x-axis to an object at position x in $[a, b]$, then the *work* W done in moving that object from position $x = a$ to $x = b$ was defined as the integral:

$$W = \int_a^b f(x)dx$$

In this section, we will see how to define the integral of a function (either realvalued or vector-valued) of two variables over a general path (i.e. a curve) in $\mathbb{R}^2$. This definition will be motivated by the physical notion of work. We will begin with real-valued functions of two variables. In physics, the intuitive idea of work is that Work = Force × Distance.

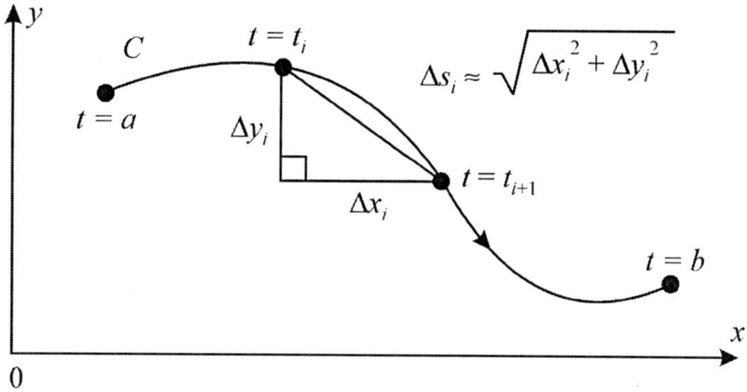

Figure: *Curve* C : x = x(t), y = y(t) *for* t *in* [a, b].

Suppose that we want to find the total amount W of work done in moving an object along a curve C in $\mathbb{R}^2$ with a smooth parametrization $x = x(t)$, $y = y(t)$, $a \leq t \leq b$, with a force $f(x, y)$ which varies with the position (x, y) of the object and is applied in the direction of motion along C.

We will assume for now that the function $f(x, y)$ is continuous and real-valued, so we only consider the magnitude of the force. Partition the interval $[a, b]$ as follows:

$$a = t_0 < t_1 < t_2 < \ldots < t_{n-1} < t_n = b, \text{ for some integer } n \geq 2$$

As we can see from Figure, over a typical subinterval $[t_i, t_{i+1}]$ the distance Δs_i traveled along the curve is approximately $\sqrt{\Delta x_i^2 + \Delta y_i^2}$, by the Pythagorean Theorem. Thus, if the subinterval is small enough than the work done in moving the object along that piece of the curve is approximately

$$\text{Force} \times \text{Distance} \approx f(x_{i^*}, y_{i^*})\sqrt{\Delta x_i^2 + \Delta y_i^2},$$

where $(x_{i*}, y_{i*}) = (x(t_{i*}), y(t_{i*}))$ for some t_{i*} in $[t_i, t_{i+1}]$, and so

$$W \approx \sum_{i=0}^{n-1} f(x_{i*}, y_{i*})\sqrt{\Delta x_i^2 + \Delta y_i^2}$$

is approximately the total amount of work done over the entire curve. But since

$$\sqrt{\Delta x_i^2 + \Delta y_i^2} = \sqrt{\left(\frac{\Delta x_i}{\Delta t_i}\right)^2 + \left(\frac{\Delta y_i}{\Delta t_i}\right)^2}\,\Delta t_i,$$

where $\Delta t_i = t_{i+1} - t_i$, then

$$W \approx \sum_{i=0}^{n-1} f(x_{i*}, y_{i*})\sqrt{\left(\frac{\Delta x_i}{\Delta t_i}\right)^2 + \left(\frac{\Delta y_i}{\Delta t_i}\right)^2}\,\Delta t_i.$$

Taking the limit of that sum as the length of the largest subinterval goes to 0, the sum over all subintervals becomes the integral from $t = a$ to $t = b$, $\frac{\Delta x_i}{\Delta t_i}$ and $\frac{\Delta y_i}{\Delta t_i}$ become $x'(t)$ and $y'(t)$, respectively, and $f(x_{i*}, y_{i*})$ becomes $f(x(t), y(t))$, so that

$$W = \int_a^b f(x(t), y(t))\sqrt{x'(t)^2 + y'(t)^2}\,dt.$$

The integral on the right side of the above equation gives us our idea of how to define, for *any* real-valued function $f(x, y)$, the integral of $f(x, y)$ along the curve C, called a *line integral*:

Definition: For a real-valued function $f(x, y)$ and a curve C in $\mathbb{R}^2$, parametrized by $x = x(t)$, $y = y(t)$, $a \le t \le b$, the line integral of $f(x, y)$ along C with respect to arc length s is

$$\int_C f(x, y)\,ds = \int_a^b f(x(t), y(t))\sqrt{x'(t)^2 + y'(t)^2}\,dt.$$

The symbol ds is the differential of the arc length function

$$s = s(t) = \int_a^t \sqrt{x'(u)^2 + y'(u)^2}\,du,$$

which you may recognize fromSection he length of the curve C over the interval $[a, t]$, for all t in $[a, b]$. That is,

$$ds = s'(t)dt = \sqrt{x'(t)^2 + y'(t)^2}\,dt,$$

by the Fundamental Theorem of Calculus. For a general real-valued function $f(x, y)$, what does the line integral $\int_C f(x,y)ds$ represent? The preceding discussion of ds gives us a clue. You can think of differentials as infinitesimal lengths. So if you think of $f(x, y)$ as the height of a picket fence along C, then $f(x, y)\, ds$ can be thought of as approximately the area of a section of that fence over some infinitesimally small section of the curve, and thus the line integral $\int_C f(x,y)ds$ is the total area of that picket fence.

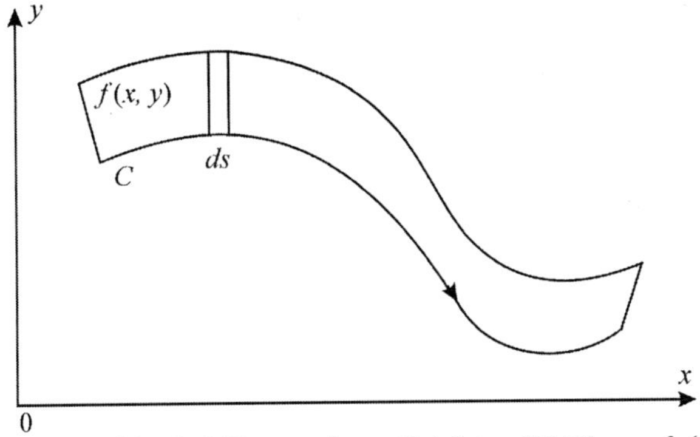

Figure: *Area of Shaded Rectangle* = *Height* × *Width* ≈ f (x, y) ds.

Example: Use a line integral to show that the lateral surface area A of a right circular cylinder of radius r and height h is $2\pi rh$.

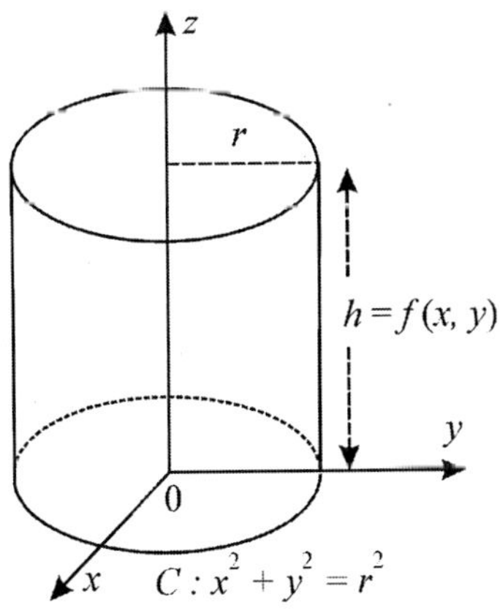

Solution: We will use the right circular cylinder with base circle C given by $x^2 + y^2 = r^2$ and with height h in the positive z direction. Parametrize C as follows:

$$x = x(t) = r \cos t,\ y = y(t) = r \sin t,\ 0 \leq t \leq 2p$$

Let $f(x, y) = h$ for all (x, y). Then

$$A = \int_C f(x,y)ds = \int_a^b f(x(t),y(t))\sqrt{x'(t)^2 + y'(t)^2}\,dt$$

$$= \int_0^{2\pi} h\sqrt{(-r\sin t)^2 + (r\cos t)^2}\,dt$$

$$= h\int_0^{2\pi} r\sqrt{\sin^2 t + \cos^2 t}\,dt$$

$$= rh\int_0^{2\pi} 1dt = 2\pi rh$$

Note in Example that if we had traversed the circle C twice, i.e. let t vary from 0 to 4π, then we would have gotten an area of $4\pi rh$, i.e. twice the desired area, even though the *curve* itself is still the same (namely, a circle of radius r). Also, notice that we traversed the circle in the counter-clockwise direction. If we had gone in the clockwise direction, using the parametrization

$$x = x(t) = r \cos(2\pi - t)\ ,\ y = y(t) = r \sin(2\pi - t)\ ,\ 0 \leq t \leq 2\pi,$$

then it is easy to verify that the value of the line integral is unchanged. In general, it can be shown that reversing the direction in which a curve C is traversed leaves $\int_C f(x,y)ds$ unchanged, for any $f(x, y)$. If a curve C has a parametrization $x = x(t)$, $y = y(t)$, $a \leq t \leq b$, then denote by $-C$ the same curve as C but traversed in the opposite direction. Then $-C$ is parametrized by

$$x = x(a + b - t)\ ,\ y = y(a + b - t)\ ,\ a \leq t \leq b,$$

and we have

$$\int_C f(x,y)ds = \int_{-C} f(x,y)ds.$$

Notice that our definition of the line integral was with respect to the arc length parameter s. We can also define

$$\int_C f(x,y)dx = \int_a^b f(x(t),y(t))x'(t)dt$$

as the *line integral of f (x, y) along C with respect to x*, and

$$\int_C f(x,y)dy = \int_a^b f(x(t),y(t))y'(t)dt$$

as the *line integral of f (x, y) along C with respect to y*.

In the derivation of the formula for a line integral, we used the idea of work as force multiplied by distance. However, we know that force is actually a *vector*. So it would be helpful to develop a vector form for a line integral. For this, suppose that we have a function $f(x, y)$ defined on $\mathbb{R}^2$ by

$$f(x, y) = P(x, y)i + Q(x, y)j$$

for some continuous real-valued functions $P(x, y)$ and $Q(x, y)$ on $\mathbb{R}^2$. Such a function f is called a vector field on $\mathbb{R}^2$. It is defined at *points* in $\mathbb{R}^2$, and its values are *vectors* in $\mathbb{R}^2$. For a curve C with a smooth parametrization $x = x(t)$, $y = y(t)$, $a \le t \le b$, let

$$r(t) = x(t)\, i + y(t)\, j$$

be the position vector for a point $(x(t), y(t))$ on C. Then $r'(t) = x'(t)\, i + y'(t)\, j$ and so

$$\int_C P(x,y)dx + \int_C Q(x,y)dy$$

$$= \int_a^b P(x(t),y(t))x'(t)dt + \int_a^b Q(x(t),y(t))y'(t)dt$$

$$= \int_a^b (P(x(t),y(t))x'(t) + Q(x(t),y(t))y'(t))dt$$

$$= \int_a^b f(x(t),y(t)).r'(t)dt$$

by definition of $f(x, y)$. Notice that the function $f(x(t), y(t)) \,.\, r'(t)$ is a *real-valued* function on $[a, b]$, so the last integral on the right looks somewhat similar to our earlier definition of a line integral. This leads us to the following definition:

Definition: For a vector field $f(x, y) = P(x, y)\, i + Q(x, y)\, j$ and a curve C with a smooth parametrization $x = x(t)$, $y = y(t)$, $a \le t \le b$, the line integral of f along C is

$$\int_C f.dr = \int_C P(x,y)dx + \int_C Q(x,y)dy$$

$$= \int_a^b f(x(t),y(t).r'(t)dt,$$

where $r(t) = x(t)\ i + y(t)\ j$ is the position vector for points on C.

We use the notation $dr = r'\ (t)dt = dx\ i + dy\ j$ to denote the differential of the vectorvalued function r. The line integral in Definition is often called a *line integral of a vector field* to distinguish it from the line integral in Definition which is called a *line integral of a scalar field*. For convenience we will often write

$$\int_C P(x,y)dx + \int_C Q(x,y)dy = \int_C P(x,y)dx + Q(x,y)dy,$$

where it is understood that the line integral along C is being applied to both P and Q. The quantity $P(x, y)\ dx + Q(x, y)\ dy$ is known as a differential form. For a realvalued function $F(x, y)$, the differential of F is $dF = \frac{\partial F}{\partial x}dx + \frac{\partial F}{\partial y}dy$. A differential form $P(x, y)\ dx + Q(x, y)\ dy$ is called exact if it equals dF for some function $F(x, y)$. Recall that if the points on a curve C have position vector $r(t) = x(t)\ i + y(t)\ j$, then $r'\ (t)$ is a tangent vector to C at the point $(x(t), y(t))$ in the direction of increasing t (which we call the *direction of* C). Since C is a smooth curve, then $r'\ (t) \neq 0$ on $[a, b]$ and hence

$$T(t) = \frac{r'(t)}{||r'(t)||}$$

is the unit tangent vector to C at $(x(t), y(t))$. Putting Definitions together we get the following theorem:

Theorem: For a vector field $f(x, y) = P(x, y)i + Q(x, y)j$ and a curve C with a smooth parametrization $x = x(t)$, $y = y(t)$, $a \leq t \leq b$ and position vector $r(t) = x(t)i + y(t)j$,

$$\int_C f.dr = \int_C f.Tds,$$

where $T(t) = \dfrac{r'(t)}{||r'(t)||}$ is the unit tangent vector to C at $(x(t), y(t))$.

If the vector field $f(x, y)$ represents the force moving an object along a curve C, then the work W done by this force is

$$W = \int_C f.Tds = \int_C f.dr.$$

Example: Evaluate $\int_C (x^2 + y^2)dx + 2xydy$, where:

(a) $C : x = t\ ,\ y = 2t\ ,\ 0 \le t \le 1$

(b) $C : x = t\ ,\ y = 2t2\ ,\ 0 \le t \le 1$

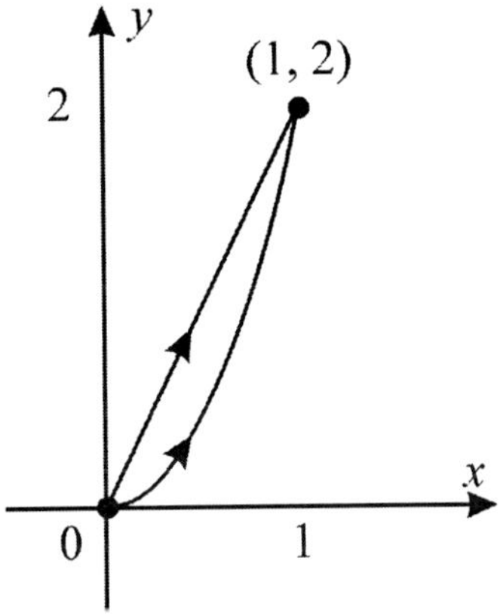

Solution: Figure shows both curves.

(a) Since $x'(t) – 1$ and $y'(t) = 2$, then

$$\int_C (x^2 + y^2)dx + 2xydy$$

$$= \int_0^1 \big((x(t)^2 + y(t)^2)x'(t) + 2x(t)y(t)y'(t)\big)dt$$

$$= \int_0^1 \big((t^2 + 4t^2)(1) + 2t(2t)(2)\big)dt$$

$$= \int_0^1 13t^2 dt$$

$$= \frac{13t^3}{3}\Bigg|_0^1 = \frac{13}{3}$$

(b) Since $x'(t) = 1$ and $y'(t) = 4t$, then

$$\int_C (x^2 + y^2)dx + 2xydy$$

$$= \int_0^1 \Big((x(t)^2 + y(t)^2)x'(t) + 2x(t)y(t)'(t)\Big)dt$$

$$= \int_0^1 \Big((t^2 + 4t^4)(1) + 2t(2t^2)(4t)\Big)dt$$

$$= \int_0^1 (t^2 + 20t^4)dt = \frac{t^3}{3} + 4t^5\Bigg|_0^1 = \frac{1}{3} + 4 = \frac{13}{3}$$

So in both cases, if the vector field $f(x, y) = (x^2 + y^2)\, i + 2xy\, j$ represents the force moving an object from (0, 0) to (1, 2) along the given curve C, then the work done is $\frac{13}{3}$. This may lead you to think that work (and more generally, the line integral of a vector field) is independent of the path taken. However, as we will see in the next section, this is not always the case. Although we defined line integrals over a single smooth curve, if C is a *piecewise smooth curve*, that is

$$C = C_1 \cup C_2 \cup .... \cup C_n$$

is the union of smooth curves C_1, ..., C_n, then we can define

$$\int_C f.dr = \int_{C_1} f.dr_1 + \int_{C_n} f.dr_2 + ...\int_{C_n} f.dr_n$$

where each ri is the position vector of the curve C_i.

Example: Evaluate $\int_C (x^2 + y^2)dx + 2xydy$, where C is the polygonal path from (0, 0) to (0, 2) to (1, 2).

Solution: Write $C = C_1 \cup C_2$, where C_1 is the curve given by $x = 0$, $y = t$, $0 \le t \le 2$ and C_2 is the curve given by $x = t$, $y = 2$, $0 \le t \le 1$. Then

$$\int_C (x^2 + y^2)dx + 2xy\,dy$$

$$= \int_{C_1} (x^2 + y^2)dx + 2xydy + \int_C (x^2 + y^2)dx + 2xy\,dy$$

$$= \int_0^2 \left((0^2 + t^2)(0) + 2(0)t(1)\right)dt +$$

$$= \int_0^2 0dt + \int_0^1 (t^2 + 4)dt$$

$$= \frac{t^3}{3} + 4t\Big|_0^1 = \frac{1}{3} + 4 = \frac{13}{3}$$

Line integral notation varies quite a bit. For example, in physics it is common to see the notation $\int_a^b f.dl$, where it is understood that the limits of integration a and b are for the underlying parameter t of the curve, and the letter l signifies length. Also, the formulation $\int_C f.Tds$ from Theorem is often preferred in physics since it emphasizes the idea of integrating the tangential component $f.T$ of f in the direction of T (i.e. in the direction of C), which is a useful physical interpretation of line integrals.

Chapter 5

Differential Calculus Basics

In mathematics, differential calculus is a subfield of calculus concerned with the study of the rates at which quantities change. It is one of the two traditional divisions of calculus, the other being integral calculus.

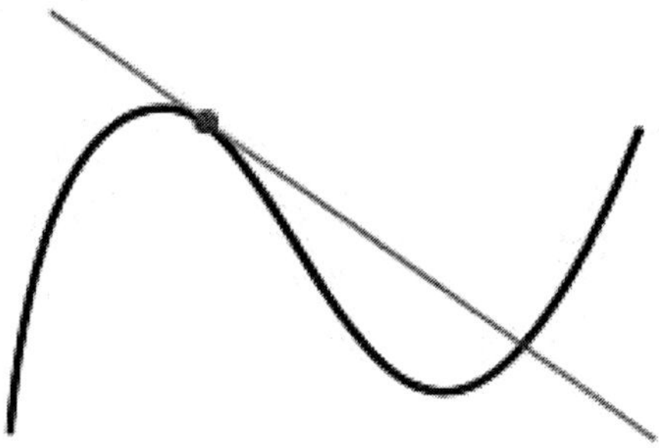

Figure: *The graph of a function, drawn in black, and a tangent line to that function, drawn in red. The slope of the tangent line equals the derivative of the function at the marked point.*

The primary objects of study in differential calculus are the derivative of a function, related notions such as the differential, and their applications. The derivative of a function at a chosen input value describes the rate of change of the function near that input value.

The process of finding a derivative is called differentiation. Geometrically, the derivative at a point equals the slope of the tangent line to the graph of the function at that point. For a real-valued function of a single real variable, the derivative of a function at a point generally determines the best linear approximation to the function at that point. Differential calculus and integral calculus are connected by the fundamental theorem of calculus, which states that differentiation is the reverse process to integration.

Differentiation has applications to nearly all quantitative disciplines. For example, in physics, the derivative of the displacement of a moving body with respect to time is the velocity of the body, and the derivative of velocity with respect to time is acceleration. Newton's second law of motion states that the derivative of the momentum of a body equals the force applied to the body.

The reaction rate of a chemical reaction is a derivative. In operations research, derivatives determine the most efficient ways to transport materials and design factories.

Derivatives are frequently used to find the maxima and minima of a function. Equations involving derivatives are called differential equations and are fundamental in describing natural phenomena. Derivatives and their generalizations appear in many fields of mathematics, such as complex analysis, functional analysis, differential geometry, measure theory and abstract algebra.

The Derivative

Suppose that x and y are real numbers and that y is a function of x, that is, for every value of x, there is a corresponding value of y.

This relationship is written as $y = f(x)$. If $f(x)$ is the equation for a straight line, then there are two real numbers m and b such that $y = m\ x + b$. m is called the slope and can be determined from the formula:

$$m = \frac{\text{change in } y}{\text{change in } x} = \frac{\Delta y}{\Delta x},$$

where the symbol Δ (the uppercase form of the Greek letter Delta) is an abbreviation for "change in". It follows that $\Delta y = m\ \Delta x$.

A general function is not a line, so it does not have a slope. The derivative of f at the point x is the best possible approximation to the idea of the slope of f at the point x.

It is usually denoted $f'(x)$ or dy/dx. Together with the value of f at x, the derivative of f determines the best linear approximation, or linearization, of f near the point x. This latter property is usually taken as the definition of the derivative.

A closely related notion is the differential of a function.

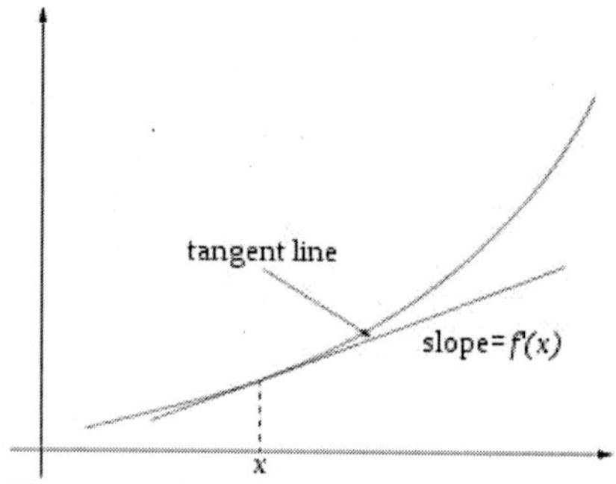

Figure: *The tangent line at (x,f(x))*

When x and y are real variables, the derivative of f at x is the slope of the tangent line to the graph of f at x. Because the source and target of f are one-dimensional, the derivative of f is a real number.

If x and y are vectors, then the best linear approximation to the graph of f depends on how f changes in several directions at once. Taking the best linear approximation in a single direction determines a partial derivative, which is usually denoted $\partial y/\partial x$. The linearization of f in all directions at once is called the total derivative.

History of Differentiation

The concept of a derivative in the sense of a tangent line is a very old one, familiar to Greek geometers such as Euclid (c. 300 BC), Archimedes (c. 287–212 BC) and Apollonius of Perga (c. 262–190 BC).

Archimedes also introduced the use of infinitesimals, although these were primarily used to study areas and volumes rather than derivatives and tangents.

The use of infinitesimals to study rates of change can be found in Indian mathematics, perhaps as early as 500 AD, when the astronomer and mathematician Aryabhata (476–550) used infinitesimals to study the motion of the moon.

The use of infinitesimals to compute rates of change was developed significantly by Bhâskara II (1114–1185); indeed, it has been argued that many of the key notions of differential calculus can be found in his work, such as "Rolle's theorem". The Persian mathematician, Sharaf al-Dîn al-Tûsî (1135–1213), was the first to discover the derivative of cubic polynomials, an important result in differential calculus; his *Treatise on Equations* developed concepts related to differential calculus, such as the derivative function and the maxima and minima of curves, in order to solve cubic equations which may not have positive solutions.

The modern development of calculus is usually credited to Isaac Newton (1643–1727) and Gottfried Leibniz (1646–1716), who provided independent and unified approaches to differentiation and derivatives.

The key insight, however, that earned them this credit, was the fundamental theorem of calculus relating differentiation and integration: this rendered obsolete most previous methods for computing areas and volumes, which had not been significantly extended since the time of Ibn al-Haytham (Alhazen).

For their ideas on derivatives, both Newton and Leibniz built on significant earlier work by mathematicians such as Isaac Barrow (1630–1677), René Descartes (1596–1650), Christiaan Huygens (1629–1695), Blaise Pascal (1623–1662) and John Wallis (1616–1703).

Isaac Barrow is generally given credit for the early development of the derivative. Nevertheless, Newton and Leibniz remain key figures in the history of differentiation, not least because Newton was the first to apply differentiation to theoretical physics, while Leibniz systematically developed much of the notation still used today.

Since the 17th century many mathematicians have contributed to the theory of differentiation. In the 19th century, calculus was put on a much more rigorous footing by mathematicians such as Augustin Louis Cauchy (1789–1857), Bernhard Riemann (1826–1866), and Karl Weierstrass (1815–1897). It was also during this period that the differentiation was generalized to Euclidean space and the complex plane.

Applications of Derivatives

Optimization: If f is a differentiable function on R (or an open interval) and x is a local maximum or a local minimum of f, then the derivative of f at x is zero; points where $f'(x) = 0$ are called *critical points* or *stationary points* (and the value of f at x is called a *critical value*). (The definition of a critical point is sometimes extended to include points where the derivative does not exist.) Conversely, a critical point x of f can be analysed by considering the second derivative of f at x:

- if it is positive, x is a local minimum;
- if it is negative, x is a local maximum;
- if it is zero, then x could be a local minimum, a local maximum, or neither. (For example, $f(x) = x^3$ has a critical point at $x = 0$, but it has neither a maximum nor a minimum there, whereas $f(x) = \pm x^4$ has a critical point at $x = 0$ and a minimum and a maximum, respectively, there.)

This is called the second derivative test. An alternative approach, called the first derivative test, involves considering the sign of the f' on each side of the critical point.

Taking derivatives and solving for critical points is therefore often a simple way to find local minima or maxima, which can be useful in optimization. By the extreme value theorem, a continuous function on a closed interval must attain its minimum and maximum values at least once. If the function is differentiable, the minima and maxima can only occur at critical points or endpoints. This also has applications in graph sketching: once the local minima and maxima of a differentiable function have been found, a rough plot of the graph can be obtained from the observation that it will be either increasing or decreasing between critical points. In higher dimensions, a critical point of a scalar valued function is a point at which the gradient is zero. The second derivative test can still be used to analyse critical points by considering the eigenvalues of the Hessian matrix of second partial derivatives of the function at the critical point. If all of the eigenvalues are positive, then the point is a local minimum; if all are negative, it is a local maximum. If there are some positive and some negative eigenvalues, then the critical point is a saddle point, and if none of these cases hold (i.e., some of the eigenvalues are zero) then the test is inconclusive.

Calculus of Variations

Calculus of variations is a field of mathematics, or more specifically calculus, that deals with maximizing or minimizing functionals, which are mappings from a set of functions to the real numbers. Functionals are often expressed as definite integrals involving functions and their derivatives. The interest is in *extremal* functions that make the functional attain a maximum or minimum value – or *stationary* functions – those where the rate of change of the functional is zero. Perhaps the simplest example of such a problem is to find the curve of shortest length, or geodesic, connecting two points.

If there are no constraints, the solution is obviously a straight line between the points. However, if the curve is constrained to lie on a surface in space, then the solution is less obvious, and possibly many solutions may exist. Such solutions are known as geodesics. A related problem is posed by Fermat's principle: light follows the path of shortest optical length connecting two points, where the optical length depends upon the material of the medium. One corresponding concept in mechanics is the principle of least action.

Many important problems involve functions of several variables. Solutions of boundary value problems for the Laplace equation satisfy the Dirichlet principle.

Plateau's problem requires finding a surface of minimal area that spans a given contour in space: the solution or solutions can often be found by dipping a wire frame in a solution of soap suds. Although such experiments are relatively easy to perform, their mathematical interpretation is far from simple: there may be more than one locally minimizing surface, and they may have non-trivial topology.

History

The calculus of variations may be said to begin with the brachistochrone curve problem raised by Johann Bernoulli (1696). It immediately occupied the attention of Jakob Bernoulli and the Marquis de l'Hôpital, but Leonhard Euler first elaborated the subject. His contributions began in 1733, and his Elementa Calculi Variationum gave to the science its name. Lagrange contributed extensively to the theory, and Legendre (1786) laid down a method, not entirely satisfactory, for the discrimination of maxima and minima. Isaac Newton and Gottfried Leibniz also gave some early attention to the subject. To this discrimination Vincenzo Brunacci (1810), Carl Friedrich Gauss (1829), Siméon Poisson (1831), Mikhail Ostrogradsky (1834), and Carl Jacobi (1837) have been among the contributors. An important general work is that of Sarrus (1842) which was condensed and improved by Cauchy (1844). Other valuable treatises and memoirs have been written by Strauch (1849), Jellett (1850), Otto Hesse (1857), Alfred Clebsch (1858), and Carll (1885), but perhaps the most important work of the century is that of Weierstrass. His celebrated course on the theory is epoch-making, and it may be asserted that he was the first to place it on a firm and unquestionable foundation.

The 20th and the 23rd Hilbert problems published in 1900 encouraged further development. In the 20th century David Hilbert, Emmy Noether, Leonida Tonelli, Henri Lebesgue and Jacques Hadamard among others made significant contributions. Marston Morse applied calculus of variations in what is now called Morse theory. Lev Pontryagin, Ralph Rockafellar and Clarke developed new mathematical tools for optimal control theory, a generalisation of calculus of variations.

Euler–Lagrange Equation

Under ideal conditions, the maxima and minima of a given function may be located by finding the points where its derivative vanishes (i.e., is equal to zero). By analogy, solutions of smooth variational problems may be obtained by solving the associated Euler–Lagrange equation.

Consider the functional

$$A[f]=\int_{x_1}^{x_2} L[x, f(x), f'(x)]dx$$

where $f'(x)=\dfrac{df}{dx}$ and where x_1 and x_2 are constants.

The function f should have at least one derivative in order to satisfy the requirements for valid application of the function; further, if the functional $A[f]$ attains its local minimum at

f_0 and $\eta(x)$ is an arbitrary function that has at least one derivative and vanishes at the endpoints x_1 and x_2, then we must have

$$A[f_0] \leq A[f_0 + \epsilon\eta]$$

for any number ε close to 0. Therefore, with $f = f_0 + \epsilon\eta$ the first variation of A must vanish,

$$\left.\frac{dA[f_0 + \epsilon\eta]}{d\epsilon}\right|_{\epsilon=0} = \int_{x_1}^{x_2} \left.\frac{dL}{d\epsilon}\right|_{\epsilon=0} dx = 0 .$$

Since L is a function of f and f', using the chain rule,

$$\frac{dL}{d\epsilon} = \frac{\partial L}{\partial f}\frac{df}{d\epsilon} + \frac{\partial L}{\partial f'}\frac{df'}{d\epsilon}$$

and since $\frac{df}{d\epsilon} = \eta$ and $\frac{df'}{d\epsilon} = \eta'$,

$$\frac{dL}{d\epsilon} = \frac{\partial L}{\partial f}\eta + \frac{\partial L}{\partial f'}\eta' .$$

Therefore,

$$\begin{aligned}
\int_{x_1}^{x_2} \left.\frac{dL}{d\epsilon}\right|_{\epsilon=0} dx &= \int_{x_1}^{x_2} \left(\frac{\partial L}{\partial f}\eta + \frac{\partial L}{\partial f'}\eta'\right) dx \\
&= \int_{x_1}^{x_2} \left(\frac{\partial L}{\partial f}\eta \quad \eta\frac{d}{dx}\frac{\partial L}{\partial f'}\right) dx + \left.\frac{\partial L}{\partial f'}\eta\right|_{x_1}^{x_2} \\
&= \int_{x_1}^{x_2} \eta\left(\frac{\partial L}{\partial f} - \frac{d}{dx}\frac{\partial L}{\partial f'}\right) dx,
\end{aligned}$$

where we have used integration by parts in the second line. The last term in the second line vanishes because $\eta \equiv 0$ at the end points. As previously mentioned the left side of the equation is zero, so according to the fundamental lemma of calculus of variations, the part of the integrand in parentheses is zero and we find that L will satisfy

$$\frac{\partial L}{\partial f}-\frac{d}{dx}\frac{\partial L}{\partial f'}=0,$$

which is called the Euler–Lagrange equation.

In general this gives a second-order ordinary differential equation which can be solved to obtain the extremal *f*. The Euler–Lagrange equation is a necessary, but not sufficient, condition for an extremal. Sufficient conditions for an extremal are discussed in the references.

In order to illustrate this process, consider the problem of finding the shortest curve in the plane that connects two points (x_1, y_1) and (x_2, y_2). The arc length is given by

$$A[f]=\int_{x_1}^{x_2}\sqrt{1+[f'(x)]^2}\,dx,$$

with

$$f'(x)=\frac{df}{dx},$$

and where $y=f(x)$, $f(x_1)=y_1$, and $f(x_2)=y_2$. Now, let $f(x)=f_0(x)+\epsilon\eta(x)$, where $f_0(x)$ is a minimizer for $A[f]$ and $\epsilon\in\mathbb{R}$ close to zero. Then

$$\frac{d}{d\epsilon}\int_{x_1}^{x_2}\sqrt{1+[f_{0'}(x)+\epsilon\eta'(x)]^2}\,dx\bigg|_{\epsilon=0}=\int_{x_1}^{x_2}\frac{(f_{0'}(x)+\epsilon\eta'(x))\eta'(x)}{\sqrt{1+[f_{0'}(x)+\epsilon\eta'(x)]^2}}\Bigg|_{\epsilon=0}$$

$$dx=\int_{x_1}^{x_2}\frac{f_{0'}(x)\eta'(x)}{\sqrt{1+[f_{0'}(x)]^2}}dx=0,$$

for any choice of the function η(though for the next step we will need to require that $\eta(x)$ vanishes at x_1, x_2). We may interpret this condition as the vanishing of all directional derivatives of $A[f_0]$ in the space of differentiable functions, and this is formalized by requiring the Fréchet derivative of *A* to vanish at f_0. If we assume that f_0 has two continuous derivatives (or if we consider weak derivatives), then we may use integration by parts:

$$\int_a^b u(x)\eta'(x)dx = [u(x)\eta(x)]_a^b - \int_a^b u'(x)\eta(x)dx$$

with the substitution

$$u(x) = \frac{f_{0'}(x)}{\sqrt{1+[f_{0'}(x)]^2}}$$

then we have

$$[u(x)\eta(x)]_{x_1}^{x_2} - \int_{x_1}^{x_2} \eta(x)\frac{d}{dx}\left[\frac{f_{0'}(x)}{\sqrt{1+[f_{0'}(x)]^2}}\right]dx = 0,$$

but the first term is zero since $\eta(x)$ was chosen to vanish at x_1 and x_2 where the evaluation is taken. Therefore,

$$\int_{x_1}^{x_2} \eta(x)\frac{d}{dx}\left[\frac{f_{0'}(x)}{\sqrt{1+[f_{0'}(x)]^2}}\right]dx = 0$$

for any twice differentiable function η that vanishes at the endpoints of the interval.

We can now apply the fundamental lemma of calculus of variations: If

$$I = \int_{x_1}^{x_2} \eta(x)H(x)dx = 0$$

for any sufficiently differentiable function $\eta(x)$ within the integration range that vanishes at the endpoints of the interval, then it follows that $H(x)$ is identically zero on its domain.

Therefore,

$$\frac{d}{dx}\left[\frac{f_{0'}(x)}{\sqrt{1+[f_{0'}(x)]^2}}\right] = 0.$$

It follows from this equation that

$$\frac{d^2 f_0}{dx^2} = 0 \Rightarrow f_0(x) = c_1 x + c_2$$

and hence the extremals are straight lines.

Beltrami Identity

In physics problems it frequently turns out that $\partial L / \partial x = 0$. In that case, the Euler-Lagrange equation can be simplified using the Beltrami identity:

$$L - f' \frac{\partial L}{\partial f'} = C, \text{ [1]}$$

where C is a constant. The left hand side is the Legendre transformation of L with respect to f'.

Du Bois Reymond's Theorem

The discussion thus far has assumed that extremal functions possess two continuous derivatives, although the existence of the integral A requires only first derivatives of trial functions. The condition that the first variation vanish at an extremal may be regarded as a weak form of the Euler-Lagrange equation.

The theorem of du Bois Reymond asserts that this weak form implies the strong form. If L has continuous first and second derivatives with respect to all of its arguments, and if

$$\frac{\partial^2 L}{(\partial f')^2} \neq 0,$$

then f_0 has two continuous derivatives, and it satisfies the Euler-Lagrange equation.

Lavrentiev Phenomenon

Hilbert was the first to give good conditions for the Euler Lagrange equations to give a stationary solution. Within a convex area and a positive thrice differentiable Lagrangian the solutions are composed of a countable collection of sections that either go along the boundary or satisfy the Euler Lagrange equations in the interior.

However Lavrentiev in 1926 showed that there are circumstances where there is no optimum solution but one can be approached arbitrarily closely by increasing numbers of sections. For instance the following:

$$L(t, x, x') = (x^3 - t)^2 x'^6,$$

$$x(0) = 0, x(1) = 1.$$

Here a zig zag path gives a better solution than any smooth path and increasing the number of sections improves the solution.

Functions of Several Variables

Variational problems that involve multiple integrals arise in numerous applications. For example, if $\varphi(x,y)$ denotes the displacement of a membrane above the domain D in the x,y plane, then its potential energy is proportional to its surface area:

$$U[\varphi] = \iint_D \sqrt{1 + \nabla\varphi \cdot \nabla\varphi}\, dx\, dy.$$

Plateau's problem consists of finding a function that minimizes the surface area while assuming prescribed values on the boundary of D; the solutions are called minimal surfaces. The Euler-Lagrange equation for this problem is nonlinear:

$$\varphi_{xx}(1 + \varphi_y^2) + \varphi_{yy}(1 + \varphi_x^2) - 2\varphi_x\varphi_y\varphi_{xy} = 0.$$

Dirichlet's Principle

It is often sufficient to consider only small displacements of the membrane, whose energy difference from no displacement is approximated by

$$V[\varphi] = \frac{1}{2}\iint_D \nabla\varphi \cdot \nabla\varphi\, dx\, dy.$$

The functional V is to be minimized among all trial functions φ that assume prescribed values on the boundary of D. If u is the minimizing function and v is an arbitrary smooth function that vanishes on the boundary of D, then the first variation of $V[u + \epsilon v]$ must vanish:

$$\frac{d}{d\epsilon} V[u + \epsilon v]\,|_{\epsilon=0} = \iint_D \nabla u \cdot \nabla v\, dx\, dy = 0.$$

Provided that u has two derivatives, we may apply the divergence theorem to obtain

$$\iint_D \nabla \cdot (v\nabla u)\, dx\, dy = \iint_D \nabla u \cdot \nabla v + v\nabla \cdot \nabla u\, dx\, dy = \iint_C v\frac{\partial u}{\partial n}\, ds,$$

where C is the boundary of D, s is arclength along C and $\partial u / \partial n$ is the normal derivative of u on C. Since υ vanishes on C and the first variation vanishes, the result is

$$\iint_D v\nabla \cdot \nabla u dxdy = 0$$

for all smooth functions v that vanish on the boundary of D. The proof for the case of one dimensional integrals may be adapted to this case to show that

$$\nabla \cdot \nabla u = 0 \text{ in } D.$$

The difficulty with this reasoning is the assumption that the minimizing function u must have two derivatives. Riemann argued that the existence of a smooth minimizing function was assured by the connection with the physical problem: membranes do indeed assume configurations with minimal potential energy. Riemann named this idea the Dirichlet principle in honour of his teacher Dirichlet. However Weierstrass gave an example of a variational problem with no solution: minimize

$$W[\varphi] = \int_{-1}^{1} (x\varphi')^2 dx$$

among all functions φ that satisfy $\varphi(-1) = -1$ and $\varphi(1) = 1$. W can be made arbitrarily small by choosing piecewise linear functions that make a transition between -1 and 1 in a small neighbourhood of the origin. However, there is no function that makes W=0. Eventually it was shown that Dirichlet's principle is valid, but it requires a sophisticated application of the regularity theory for elliptic partial differential equations.

Generalization to Other Boundary Value Problems

A more general expression for the potential energy of a membrane is

$$V[\varphi] = \iint_D \left[\frac{1}{2}\nabla\varphi \cdot \nabla\varphi + f(x,y)\varphi\right] dxdy + \int_C \left[\frac{1}{2}\sigma(s)\varphi^2 + g(s)\varphi\right] ds.$$

This corresponds to an external force density $f(x, y)$ in D, an external force $g(s)$ on the boundary C, and elastic forces with modulus $\sigma(s)$ acting on C.

The function that minimizes the potential energy with no restriction on its boundary values will be denoted by u. Provided that f and g are continuous, regularity theory implies that the minimizing function u will have two derivatives. In taking the first variation, no boundary condition need be imposed on the increment v. The first variation of $V[u+\epsilon v]$ is given by

$$\iint_D [\nabla u \cdot \nabla v + fv]dxdy + \int_C [\sigma uv + gv]ds = 0.$$

If we apply the divergence theorem, the result is

$$\iint_D [-v\nabla \cdot \nabla u + vf]dxdy + \int_C v\left[\frac{\partial u}{\partial n} + \sigma u + g\right]ds = 0.$$

If we first set v=0 on C, the boundary integral vanishes, and we conclude as before that

$$-\nabla \cdot \nabla u + f = 0$$

in D. Then if we allow v to assume arbitrary boundary values, this implies that u must satisfy the boundary condition

$$\frac{\partial u}{\partial n} + \sigma u + g = 0,$$

on C. Note that this boundary condition is a consequence of the minimizing property of u: it is not imposed beforehand. Such conditions are called natural boundary conditions.

The preceding reasoning is not valid if σ vanishes identically on C. In such a case, we could allow a trial function $\varphi \equiv c$, where c is a constant. For such a trial function,

$$V[c] = c\left[\iint_D f\,dxdy + \int_C gds\right].$$

By appropriate choice of c, V can assume any value unless the quantity inside the brackets vanishes. Therefore the variational problem is meaningless unless

$$\iint_D f\,dxdy + \int_C gds = 0.$$

This condition implies that net external forces on the system are in equilibrium.

If these forces are in equilibrium, then the variational problem has a solution, but it is not unique, since an arbitrary constant may be added. Further details and examples are in Courant and Hilbert (1953).

Eigenvalue Problems

Both one-dimensional and multi-dimensional eigenvalue problems can be formulated as variational problems.

Sturm-Liouville Problems

The Sturm-Liouville eigenvalue problem involves a general quadratic form

$$Q[\varphi] = \int_{x_1}^{x_2} \left[p(x)\varphi'(x)^2 + q(x)\varphi(x)^2 \right] dx,$$

where φ is restricted to functions that satisfy the boundary conditions

$$\varphi(x_1) = 0, \quad \varphi(x_2) = 0.$$

Let R be a normalization integral

$$R[\varphi] = \int_{x_1}^{x_2} r(x)\varphi(x)^2 \, dx.$$

The functions $p(x)$ and $r(x)$ are required to be everywhere positive and bounded away from zero. The primary variational problem is to minimize the ratio Q/R among all φ satisfying the endpoint conditions. It is shown below that the Euler-Lagrange equation for the minimizing u is

$$-(pu')' + qu - \lambda ru = 0,$$

where λ is the quotient

$$\lambda = \frac{Q[u]}{R[u]}.$$

It can be shown that the minimizing u has two derivatives and satisfies the Euler-Lagrange equation. The associated λ will be denoted by λ_1; it is the lowest eigenvalue for this equation and boundary conditions. The associated minimizing function will be denoted by $u_1(x)$. This variational characterization of eigenvalues leads to the Rayleigh-Ritz method: choose an approximating u as a linear combination of basis functions (for example trigonometric functions) and carry out a finite-dimensional minimization among such linear combinations. This method is often surprisingly accurate.

The next smallest eigenvalue and eigenfunction can be obtained by minimizing Q under the additional constraint

$$\int_{x_1}^{x_2} r(x)u_1(x)\varphi(x)dx = 0.$$

This procedure can be extended to obtain the complete sequence of eigenvalues and eigenfunctions for the problem.

The variational problem also applies to more general boundary conditions.

Instead of requiring that φ vanish at the endpoints, we may not impose any condition at the endpoints, and set

$$Q[\varphi] = \int_{x_1}^{x_2} \left[p(x)\varphi'(x)^2 + q(x)\varphi(x)^2 \right] dx + a_1\varphi(x_1)^2 + a_2\varphi(x_2)^2,$$

where a_1 and a_2 are arbitrary. If we set $\varphi = u + \epsilon v$ the first variation for the ratio Q / R is

$$V_1 = \frac{2}{R[u]} \left(\begin{array}{l} \int_{x_1}^{x_2} [p(x)u'(x)v'(x) + q(x)u(x)v(x) - \lambda u(x)v(x)] \\ dx + a_1 u(x_1)v(x_1) + a_2 u(x_2)v(x_2) \end{array} \right),$$

where λ is given by the ratio $Q[u] / R[u]$ as previously. After integration by parts,

$$\frac{R[u]}{2} V_1 = \int_{x_1}^{x_2} v(x)\left[-(pu')' + qu - \lambda ru \right] dx + v(x_1)[-p(x_1)u'(x_1)$$

$$+a_1 u(x_1)] + v(x_2)[p(x_2)u'(x_2) + a_2 u(x_2)].$$

If we first require that v vanish at the endpoints, the first variation will vanish for all such v only if

$$-(pu')' + qu - \lambda ru = 0 \quad \text{for} \quad x_1 < x < x_2.$$

If u satisfies this condition, then the first variation will vanish for arbitrary v only if

$$-p(x_1)u'(x_1) + a_1 u(x_1) = 0, \quad \text{and} \quad p(x_2)u'(x_2) + a_2 u(x_2) = 0.$$

These latter conditions are the natural boundary conditions for this problem, since they are not imposed on trial functions for the minimization, but are instead a consequence of the minimization.

Eigenvalue Problems in Several Dimensions

Eigenvalue problems in higher dimensions are defined in analogy with the one-dimensional case.

For example, given a domain D with boundary B in three dimensions we may define

$$Q[\varphi] = \iiint_D p(X)\nabla\varphi\cdot\nabla\varphi + q(X)\varphi^2\,dx\,dy\,dz + \iint_B \sigma(S)\varphi^2\,dS,$$

and $R[\varphi] = \iiint_D r(X)\varphi(X)^2\,dx\,dy\,dz.$

Let u be the function that minimizes the quotient $Q[\varphi]/R[\varphi]$, with no condition prescribed on the boundary B. The Euler-Lagrange equation satisfied by u is

$$-\nabla\cdot(p(X)\nabla u) + q(x)u - \lambda r(x)u = 0,$$

where

$$\lambda = \frac{Q[u]}{R[u]}.$$

The minimizing u must also satisfy the natural boundary condition

$$p(S)\frac{\partial u}{\partial n} + \sigma(S)u = 0,$$

on the boundary B. This result depends upon the regularity theory for elliptic partial differential equations.

Many extensions, including completeness results, asymptotic properties of the eigenvalues and results concerning the nodes of the eigenfunctions are in Courant and Hilbert (1953).

Applications

Some applications of the Calculus of variations include:

- The derivation of the Catenary shape
- The Brachistochrone problem
- Isoperimetric problems
- Geodesics on surfaces
- Minimal surfaces and Plateau's problem
- Optimal Control.

Fermat's Principle

Fermat's principle states that light takes a path that (locally) minimizes the optical length between its endpoints. If the x-coordinate is chosen as the parameter along the path, and $y = f(x)$ along the path, then the optical length is given by

$$A[f] = \int_{x=x_0}^{x_1} n(x, f(x))\sqrt{1 + f'(x)^2}\, dx,$$

where the refractive index $n(x, y)$ depends upon the material. If we try $f(x) = f_0(x) + \epsilon f_1(x)$ then the first variation of A (the derivative of A with respect to ε) is

$$\delta A[f_0, f_1] = \int_{x=x_0}^{x_1} \left[\frac{n(x, f_0) f_{0'}(x) f_{1'}(x)}{\sqrt{1 + f_{0'}(x)^2}} + n_y(x, f_0) f_1 \sqrt{1 + f_{0'}(x)^2} \right] dx.$$

After integration by parts of the first term within brackets, we obtain the Euler-Lagrange equation

$$-\frac{d}{dx}\left[\frac{n(x, f_0) f_{0'}}{\sqrt{1 + f_{0'}^2}} \right] + n_y(x, f_0)\sqrt{1 + f_{0'}(x)^2} = 0.$$

The light rays may be determined by integrating this equation. This formalism is used in the context of Lagrangian optics and Hamiltonian optics.

Snell's Law

There is a discontinuity of the refractive index when light enters or leaves a lens. Let

$$n(x, y) = n_- \quad \text{if} \quad x < 0,$$

$$n(x, y) = n_+ \quad \text{if} \quad x > 0,$$

where n_- and n_+ are constants. Then the Euler-Lagrange equation holds as before in the region where $x<0$ or $x>0$, and in fact the path is a straight line there, since the refractive index is constant. At the $x=0$, f must be continuous, but f' may be discontinuous. After integration by parts in the separate regions and using the Euler-Lagrange equations, the first variation takes the form

$$\delta A[f_0, f_1] = f_1(0)\left[n_- \frac{f_{0'}(0_-)}{\sqrt{1+f_{0'}(0_-)^2}} - n_+ \frac{f_{0'}(0_+)}{\sqrt{1+f_{0'}(0_+)^2}} \right].$$

The factor multiplying n_- is the sine of angle of the incident ray with the x axis, and the factor multiplying n_+ is the sine of angle of the refracted ray with the x axis. Snell's law for refraction requires that these terms be equal. As this calculation demonstrates, Snell's law is equivalent to vanishing of the first variation of the optical path length.

Fermat's Principle in Three Dimensions

It is expedient to use vector notation: let $X = (x_1, x_2, x_3)$, let t be a parameter, let $X(t)$ be the parametric representation of a curve C, and let $\dot{X}(t)$ be its tangent vector. The optical length of the curve is given by

$$A[C] = \int_{t=t_0}^{t_1} n(X)\sqrt{\dot{X}\cdot\dot{X}}\,dt.$$

Note that this integral is invariant with respect to changes in the parametric representation of C. The Euler-Lagrange equations for a minimizing curve have the symmetric form

$$\frac{d}{dt}P = \sqrt{\dot{X}\cdot\dot{X}}\,\nabla n,$$

where

$$P = \frac{n(X)\dot{X}}{\sqrt{\dot{X}\cdot\dot{X}}}.$$

It follows from the definition that P satisfies

$$P\cdot P = n(X)^2.$$

Therefore the integral may also be written as

$$A[C] = \int_{t=t_0}^{t_1} P \cdot \dot{X}\, dt.$$

This form suggests that if we can find a function ψ whose gradient is given by *P*, then the integral *A* is given by the difference of ψ at the endpoints of the interval of integration. Thus the problem of studying the curves that make the integral stationary can be related to the study of the level surfaces of ψ. In order to find such a function, we turn to the wave equation, which governs the propagation of light.

This formalism is used in the context of Lagrangian optics and Hamiltonian optics.

Connection with the Wave Equation

The wave equation for an inhomogeneous medium is

$$u_{tt} = c^2 \nabla \cdot \nabla u,$$

where *c* is the velocity, which generally depends upon *X*. Wave fronts for light are characteristic surfaces for this partial differential equation: they satisfy

$$\varphi_t^2 = c(X)^2 \nabla \varphi \cdot \nabla \varphi.$$

We may look for solutions in the form

$$\varphi(t, X) = t - \psi(X).$$

In that case, ψ satisfies

$$\nabla \psi \cdot \nabla \psi = n^2,$$

where $n = 1/c$. According to the theory of first-order partial differential equations, if $P = \nabla \psi$, then *P* satisfies

$$\frac{dP}{ds} = n \nabla n,$$

along a system of curves (the light rays) that are given by

$$\frac{dX}{ds} = P.$$

These equations for solution of a first-order partial differential equation are identical to the Euler-Lagrange equations if we make the identification

$$\frac{ds}{dt} = \frac{\sqrt{\dot{X}\cdot\dot{X}}}{n}.$$

We conclude that the function ψ is the value of the minimizing integral A as a function of the upper end point. That is, when a family of minimizing curves is constructed, the values of the optical length satisfy the characteristic equation corresponding the wave equation. Hence, solving the associated partial differential equation of first order is equivalent to finding families of solutions of the variational problem. This is the essential content of the Hamilton-Jacobi theory, which applies to more general variational problems.

Action Principle

In classical mechanics, the action, S, is defined as the time integral of the Lagrangian, L. The Lagrangian is the difference of energies,

$$L = T - U,$$

where T is the kinetic energy of a mechanical system and U its potential energy. Hamilton's principle (or the action principle) states that the motion of a conservative holonomic (integrable constraints) mechanical system is such that the action integral

$$S = \int_{t=t_0}^{t_1} L(x, \dot{x}, t)dt$$

is stationary with respect to variations in the path *x(t)*. The Euler-Lagrange equations for this system are known as Lagrange's equations:

$$\frac{d}{dt}\frac{\partial L}{\partial \dot{x}} = \frac{\partial L}{\partial x},$$

and they are equivalent to Newton's equations of motion (for such systems).

The conjugate momenta P are defined by

$$p = \frac{\partial L}{\partial \dot{x}}.$$

For example, if

$$T = \frac{1}{2} m\dot{x}^2,$$

then

$$p = m\dot{x}.$$

Hamiltonian mechanics results if the conjugate momenta are introduced in place of *x*, and the Lagrangian *L* is replaced by the Hamiltonian *H* defined by

$$H(x, p, t) = p\dot{x} - L(x, \dot{x}, t)$$

The Hamiltonian is the total energy of the system: $H = T + U$. Analogy with Fermat's principle suggests that solutions of Lagrange's equations (the particle trajectories) may be described in terms of level surfaces of some function of *X*. This function is a solution of the Hamilton-Jacobi equation:

$$\frac{\partial \psi}{\partial t} + H\left(x, \frac{\partial \psi}{\partial x}, t\right) = 0.$$

Physics

Calculus is of vital importance in physics: many physical processes are described by equations involving derivatives, called differential equations. Physics is particularly concerned with the way quantities change and evolve over time, and the concept of the "time derivative" — the rate of change over time — is essential for the precise definition of several important concepts. In particular, the time derivatives of an object's position are significant in Newtonian physics:

- velocity is the derivative (with respect to time) of an object's displacement (distance from the original position)
- acceleration is the derivative (with respect to time) of an object's velocity, that is, the second derivative (with respect to time) of an object's position.

Chapter 6

Linear Differential Equation

In mathematics, a linear differential equation is a differential equation of the form

$$Ly = f$$

where the differential operator L is a linear operator, y is the unknown function, and the right hand side f is a given function (called the source term). The linearity condition on L rules out operations such as taking the square of the derivative of y; but permits, for example, taking the second derivative of y. Therefore a fairly general form of such an equation would be

$$a_n(x)\, D^n\, y(x) + a_{n-1}\,(x)\, D^{n-1}\, y(x)$$
$$+\ldots + a_1(x)\, Dy(x) + a_0(x)y(x) = f(x)$$

where D is the differential operator d/dx (i.e. $Dy = y'$, $D^2y = y''$,..), and the a_i are given functions. Such an equation is said to have order n, the index of the highest derivative of y that is involved. (Assuming a possibly existing coefficient a_n of this derivative to be non zero, it is eliminated by dividing through it.

In case it can become zero, different cases must be considered separately for the analysis of the equation.) If y is assumed to be a function of only one variable, one speaks about an ordinary differential equation, else the derivatives and their coefficients must be understood as (contracted) vectors, matrices or tensors of higher rank, and we have a (linear) partial differential equation. The case where $f = 0$ is called a homogeneous equation and its solutions are called complementary functions. It is particularly important to the solution of the general case, since any complementary function can be added to a solution of the inhomogeneous equation to give another solution (by a method traditionally called *particular integral and complementary function*). When the a_i are numbers, the equation is said to have *constant coefficients.*

The subject of ordinary differential equations encompasses such a large field that you can make a profession of it. There are however a small number of techniques in the subject that you have to know. These are the ones that come up so often in physical systems that you need not only the skills to use them but the intuition about what they will do.

Linear Constant-Coefficient

A differential equation such as

$$\left(\frac{d^2x}{dt^2}\right)^3 + t^2x^4 + 1 = 0$$

is difficult to solve, sinceit is non-linear. This means that if I have two solutions $x_1(t)$ and $x_2(t)$ then the sum $x_1 + x_2$ is not a solution; look at all the cross-terms you get if you try to plug the sum in to the equation. Also if you multiply $x_1(t)$ by 2 you no longer have a solution.

An equation such as

$$e^t\frac{d^3x}{dt^3} + t^2\frac{dx}{dt} - x = 0$$

may be a mess to solve, but if you have two solutions, $x_1(t)$ and $x_2(t)$ then the sum $\alpha x_1 + \beta x_2$ is also a solution. This is called a linear, homogeneous equation because of this property. A similar-looking equation,

$$e^t\frac{d^3x}{dt^3} + t^2\frac{dx}{dt} - x = t$$

does not have this property, though it's close. It's called a linear, inhomogeneous equation. If $x_1(t)$ and $x_2(t)$ are solutions to this, then if you try their sum as a solution you get $2t = t$, and that's no solution, but it misses working only because of the single term on the right. One of the most common sorts of differential equations that you see is an especially simple one to solve. That's part of the reason it's so common.

This is the linear, constant-coefficient, differential equation. If you have a mass tied to the end of a spring and the other end of the spring is fixed, the force applied to the mass by the spring is to a good approximation proportional to the distance that the mass has moved from its equilibrium position.

If the coordinate x is measured from the mass's equilibrium position, the equation $\vec{F} = m\vec{a}$ says

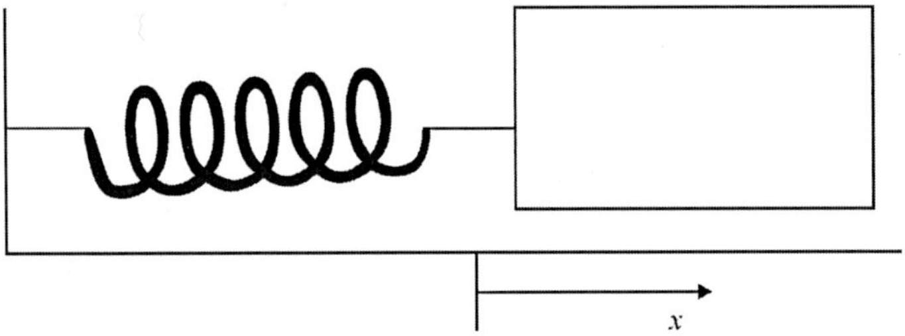

$$m\frac{d^2x}{dt^2} = -kx$$

If there's friction (and there's *always* friction), the force has another term. Now how do you describe friction mathematically? The common model for dry friction is that the magnitude of the force is independent of the magnitude of the mass's velocity and opposite to the direction of the velocity. If you try to write that down in a compact mathematical form you get something like,

$$\vec{F}_{friction} = -\mu_k F_N \frac{\vec{v}}{|\vec{v}|}$$

This is hard to work with. It can be done, but you are going to do something different. Wet friction is easier to handle mathematically because when you lubricate a surface, the friction becomes velocity dependent in a way that is, for low speeds, proportional to the velocity.

Neither of these two representations is a completely accurate description of the way friction works. That's far more complex than either of these simple models, but these approximations are good enough for many purposes and I'll settle for them. Assume "wet friction" and the differential equation for the motion of m is

$$m\frac{d^2x}{dt^2} = -kx - b\frac{dx}{dt}$$

This is a second order, linear, homogeneous, differential equation, which simply means that the highest derivative present is the second, the sum of two solutions is a solution, and a constant multiple of a solution is a solution.

That the coefficients are constants makes this an easy equation to solve. All you have to do is to recall that the derivative of an exponential is an exponential. $de^t/dt = e^t$. If you substitute this exponential for $x(t)$, of course it can't work as a solution; it doesn't even make sense dimensionally. What is e to the power of a day?

You have to have something in the exponent to make it dimensionless, $e^{\alpha t}$.

Also, the function x is supposed to give you a position, with dimensions of length. Use another constant: $x(t) = Ae^{\alpha t}$. Plug *this* into the differential equation to find

$$mA\alpha^2 e^{\alpha t} + bA\alpha e^{\alpha t} + kAe^{\alpha t} = Ae^{\alpha t}[m\alpha^2 + b\alpha + k] = 0$$

The product of factors is zero, and the only way that a product of two numbers can be zero is if one of the numbers is zero. The exponential never vanishes, and for a non-trivial solution $A \neq 0$, so all that's left is the polynomial in α.

$$m\alpha^2 + b\alpha + k = 0,$$

with solutions $\alpha = \dfrac{-b \pm \sqrt{b^2 - 4km}}{2m}$

The position function is then

$$x(t) = Ae^{\alpha_1 t} + Be^{\alpha_2 t}$$

where A and B are arbitrary constants and α_1 and α_2 are the two roots. Supposed that it is oscillating? It is a harmonic oscillator after all, but the exponentials don't look very oscillatory. If you have a mass on the end of a spring and the entire system is immersed in honey, it won't do much oscillating! Translated into mathematics, this says that if the constant b is too large, there is no oscillation.

In the equation for α, if b is large enough the argument of the square root is positive and both α's are real — no oscillation. Only if b is small enough does the argument of the square root become negative; then you get complex values for the α's and hence oscillations. Push this to the extreme case where the damping vanishes: $b = 0$.

Then $\alpha_1 = i\sqrt{k/m}$ and $\alpha_2 = -i\sqrt{k/m}$.

Denote $\omega_0 = \sqrt{k/m}$.

$$x(t) = Ae^{i\omega_0 t} + Be^{-i\omega_0 t}$$

You can write this in other forms, see problem 10. To determine the arbitrary constant A and B you need two equations. They come from some additional information about the problem, typically some initial conditions. Take a specific example in which you start from the origin with a kick, $x(0) = 0$ and $\dot{x}(0) = v_0$.

$$x(0) = 0 = A + B, \quad \dot{x}(0) = v0 = \mathrm{i}\,\omega_0\,\mathrm{A} - \omega_0\,\mathrm{B}$$

Solve for A and B to get $A = -B = v_0/(2\mathrm{i}\,\omega_0)$. Then

$$x(t) = \frac{v_0}{2i\omega_0}[e^{i\omega_0 t} - e^{-i\omega_0 t}] = \frac{v_0}{\omega_0}\sin\omega_0 t$$

As a check on the algebra, use the first term in the power series expansion of the sine function to see how x behaves for small t. The sine factor is $\sin\,\omega_0 t \approx \omega_0 t$, and then $x(t)$ is approximately $v_0 t$, just as it should be. Also notice that despite all the complex numbers, the final answer is real.

Damped Oscillator

If there is damping, but not too much, then the α's have an imaginary part *and* a negative real part.

$$\alpha = \frac{-b \pm i\sqrt{4km - b^2}}{2m} = -\frac{b}{2m} \pm i\omega', \text{ where}$$

$$\omega' = \sqrt{\frac{k}{m} - \frac{b^2}{4m^2}}$$

This represents a damped oscillation and has frequency a bit lower than the one in the undamped case. Use the same initial conditions as above and you will get similar results (let $(\gamma = b/2m)$

$$x(t) = Ae^{(-\gamma + i\omega')t} + Be^{(-\gamma - i\omega')t}$$

$$x(0) = A + B = 0,\ v_x(0) = (-\gamma + i\omega')A + (-\gamma - i\omega')B = v_0$$

The two equations for the unknowns A and B imply $B = -A$ and

$$2i\omega' A = v_0, \text{ so } x(t) = \frac{v_0}{2i\omega'} e^{-\gamma t}[e^{i\omega' t} - e^{-i\omega' t}] \frac{v_0}{\omega'} e^{-\gamma t} \sin \omega' t$$

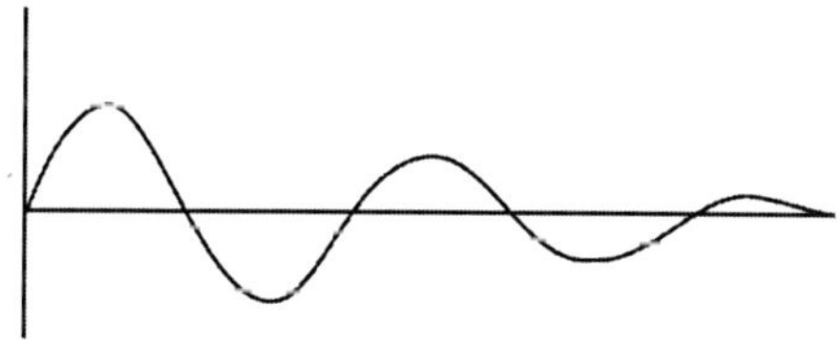

For small values of t, the first terms in the power series expansion of this result are

$$x(t) = \frac{v_0}{\omega'}[1 - \gamma t + \gamma^2 t^2 / 2 - ...]$$

$$[\omega' t - \omega'^3 t^3 / 6 + ...] = v_0 t - v_0 \gamma t^2 + ...$$

The first term is what you should expect, as the initial velocity is $v_x = v_0$. The negative sign in the next term says that it doesn't move as far as it would without the damping, but analyze it further. Does it have the right size as well as the right sign? It is $-v_0\gamma t^2 = -v_0(b/2m)t^2$. But that's an acceleration: $a_x t^2/2$. It says that the acceleration just after the motion starts is $a_x = -bv_0/m$. Is that what you should expect? As the motion starts, the mass hasn't gone very far so the spring doesn't yet exert much force. The viscous friction is however $-bv_x$. Set that equal to ma_x and you see that,

$$-v_0\gamma t^2$$

has precisely the right value. In Equation you assumed that the two roots of the quadratic, the two α's, are different. What if they aren't? Then you have only one value of to use in defining the solution $e^{\alpha t}$ in Equation. You now have only one arbitrary constant with which to match two initial conditions. What is the energy for this damped oscillator? The kinetic energy is $mv^2/2$ and the potential energy for the spring is $kx^2/2$. Is the sum constant? No.

$$\frac{d}{dt}\frac{1}{2}(mv^2 + kx^2) = mv\frac{dv}{dt} + kx\frac{dx}{dt}$$

$$= v_x(ma_x + kx) = -bv_x^2 = F_{x},\text{frict } v_x$$

"Force times velocity" is a common expression for power, and this says that the total energy is decreasing according to this formula. The energy decreases exponentially on average.

Sequences of Functions

Let's start off this section with a discussion of just what a sequence is. A sequence is nothing more than a list of numbers written in a specific order. The list may or may not have an infinite number of terms in them although we will be dealing exclusively with infinite sequences in this class. General sequence terms are denoted as follows,

$$a_1 - \text{first term}$$
$$a_2 - \text{second term}$$
$$\vdots$$
$$a_n - n^{th} \text{ term}$$
$$a_{n+1} - (n+1)^{st} \text{ term}$$
$$\vdots$$

Because we will be dealing with infinite sequences each term in the sequence will be followed by another term as noted above. In the notation above we need to be very careful with the subscripts. The subscript of $n + 1$ denotes the next term in the sequence and NOT one plus the nth term! In other words,

$$a_{n+1} \neq a_n + 1$$

So be very careful when writing subscripts to make sure that the "+1" doesn't migrate out of the subscript! This is an easy mistake to make when you first start dealing with this kind of thing.

There is a variety of ways of denoting a sequence. Each of the following are equivalent ways of denoting a sequence.

$$\{a_1, a_2, \ldots, a_n, a_{n+1}, \ldots\} \qquad \{a_n\} \qquad \{a_n\}_{n+1}^{\infty}$$

In the second and third notations above a_n is usually given by a formula. A couple of notes are now in order about these notations. First, note the difference between the second and third notations above. If the starting point is not important or is implied in some way by the problem it is often not written down as we did in the third notation. Next, we used a starting point of $n = 1$ in the third notation only so we could write one down. There is absolutely no reason to believe that a sequence will start at $n = 1$. A sequence will start where ever it needs to start.

Let's take a look at a couple of sequences.

Example: Write down the first few terms of each of the following sequences.

1. $\left\{\dfrac{n+1}{n^2}\right\}_{n=1}^{\infty}$

2. $\left\{\dfrac{(-1)^{n+1}}{2^n}\right\}_{n=0}^{\infty}$

3. $\left\{b_n\right\}_{n=0}^{\infty}$, where $b_n = n^{th}$ digit of π

Solution:

1. $\left\{\frac{n+1}{n^2}\right\}_{n=1}^{\infty}$

To get the first few sequence terms here all we need to do is plug in values of n into the formula given and we'll get the sequence terms.

$$\left\{\frac{n+1}{n^2}\right\}_{n=1}^{\infty} = \left\{\underbrace{2}_{n=1}, \underbrace{\frac{3}{4}}_{n=2}, \underbrace{\frac{4}{9}}_{n=3}, \underbrace{\frac{5}{16}}_{n=4}, \underbrace{\frac{6}{25}}_{n=5}, \ldots\right\}$$

Note the inclusion of the "..." at the end! This is an important piece of notation as it is the only thing that tells us that the sequence continues on and doesn't terminate at the last term.

2. $\left\{\frac{(-1)^{n+1}}{2^n}\right\}_{n=0}^{\infty}$

This one is similar to the first one. The main difference is that this sequence doesn't start at $n = 1$.

$$\left\{\frac{(-1)^{n+1}}{2^n}\right\}_{n=0}^{\infty} = \left\{-1, \frac{1}{2}, \frac{1}{4}, \frac{1}{8}, \frac{1}{16}, \ldots\right\}$$

Note that the terms in this sequence alternate in signs. Sequences of this kind are sometimes called alternating sequences.

3. $\left\{b_n\right\}_{n=1}^{\infty}$, where $b_n = n^{th}$ digit of π

This sequence is different from the first two in the sense that it doesn't have a specific formula for each term. However, it does tell us what each term should be.

Each term should be the nth digit of π. So we know that p = 3.14159265359...

The sequence is then,

$$\{3, 1, 4, 1, 5, 9, 2, 6, 5, 3, 5,...\}$$

In the first two parts of the previous example note that we were really treating the formulas as functions that can only have integers plugged into them. Or,

$$f(n) = \frac{n+1}{n^2} \qquad g(n) = \frac{(-1)^{n+1}}{2^n}$$

This is an important idea in the study of sequences (and series). Treating the sequence terms as function evaluations will allow us to do many things with sequences that couldn't do otherwise. Before delving further into this idea however we need to get a couple more ideas out of the way. First we want to think about "graphing" a sequence. To graph the sequence $\{a_n\}$ we plot the points $(n_3\ a_n)$ as n ranges over all possible values on a graph.

For instance, let's graph the sequence $\left\{\frac{n+1}{n^2}\right\}_{n=1}^{\infty}$. The first few points on the graph are,

$$(1, 2), \left(2, \frac{3}{3}\right), \left(3, \frac{4}{9}\right), \left(4, \frac{5}{16}\right), \left(5, \frac{6}{25}\right), ...$$

The graph, for the first 30 terms of the sequence, is then,

This graph leads us to an important idea about sequences. Notice that as n increases the sequence terms from our sequence terms, in this case, get closer and closer to zero. We then say that zero is the limit (or sometimes the limiting value) of the sequence and write,

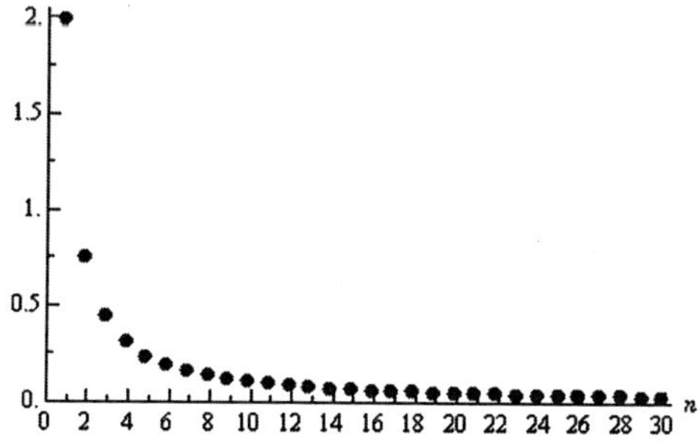

$$\lim_{n\to\infty} a_n = \lim_{n\to\infty} \frac{n+1}{n^2} = 0$$

This notation should look familiar to you. It is the same notation we used when we talked about the limit of a function. In fact, if you recall, we said earlier that we could think of sequences as functions in some way and so this notation shouldn't be too surprising.

Using the ideas that we developed for limits of functions we can write down the following *working definition* for limits of sequences.

Working Definition of Limit

1. We say that

   $$\lim_{n\to\infty} a_n = L$$

   if we can make an as close to L as we want for all sufficiently large n. In other words, the value of the a_n's approach L as n approaches infinity.

2. We say that

   $$\lim_{n\to\infty} a_n = \infty$$

   if we can make a_n as large as we want for all sufficiently large n. Again, in other words, the value of the a_n's get larger and larger without bound as n approaches infinity.

3. We say that

   $$\lim_{n\to\infty} a_n = -\infty$$

If we can make an as large and negative as we want for all sufficiently large n.

Again, in other words, the value of the a_n's are negative and get larger and larger without bound as n approaches infinity.

The working definitions of the various sequence limits are nice in that they help us to visualize what the limit actually is. Just like with limits of functions however, there is also a precise definition for each of these limits. Let's give those before proceeding.

Precise Definition of Limit

1. We say that $\lim_{n\to\infty} a_n = L$ if for every number Ý > 0 there is an integer N such that

   $| an - L | <$ ⥣ whenever $n > N$

2. We say that $\lim_{n\to\infty} a_n = \infty$ if for every number $M > 0$ there is an integer N such that

   $an > M$ whenever $n > N$

3. We say that $\lim_{n\to\infty} a_n$ if for every number $M < 0$ there is an integer N such that

   $an < M$ whenever $n > N$

We won't be using the precise definition often, but it will show up occasionally.

Note that both definitions tell us that in order for a limit to exist and have a finite value all the sequence terms must be getting closer and closer to that finite value as n increases. Now that we have the definitions of the limit of sequences out of the way we have a bit of terminology that we need to look at.

If $\lim_{n\to\infty} a_n$ exists and is finite we say that the sequence is convergent. If $\lim_{n\to\infty} a_n$ doesn't exist or is infinite we say the sequence diverges. Note that sometimes we will say the sequence diverges to $-\infty$ if $\lim_{n\to\infty} a_n = \infty$ and if $\lim_{n\to\infty} a_n$ we will sometimes say that the sequence diverges to $-\infty$.

Get used to the terms "convergent" and "divergent" as we'll be seeing them quite a bit throughout this chapter.

So just how do we find the limits of sequences? Most limits of most sequences can be found using one of the following theorems.

Theorem: Given the sequence $\{a_n\}$ if we have a function $f(x)$ such that $f(n) = a_n$ an $\lim_{n\to\infty} f(x) = L$ then $\lim_{n\to\infty} a_n = L$

This theorem is basically telling us that we take the limits of sequences much like we take the limit of functions. In fact, in most cases we'll not even really use this theorem by explicitly writing down a function. We will more often just treat the limit as if it were a limit of a function and take the limit as we always did back in Calculus I when we were taking the limits of functions.

So, now that we know that taking the limit of a sequence is nearly identical to taking the limit of a function we also know that all the properties from the limits of functions will also hold.

Properties

1. $\lim_{n\to\infty}(a_n \pm b_n) = \lim_{n\to\infty} a_n \pm \lim_{n\to\infty} b_n$

2. $\lim_{n\to\infty} ca_n = c \lim_{n\to\infty} a_n$

3. $\lim_{n\to\infty}(a_n b_n) = \left(\lim_{n\to\infty} a_n\right)\left(\lim_{n\to\infty} b_n\right)$

4. $\lim_{n\to\infty} \frac{a_n}{b_n} = \frac{\lim_{n\to\infty} a_n}{\lim_{n\to\infty} b_n}$, provided $\lim_{n\to\infty} b_n \neq 0$

5. $\lim_{n\to\infty} a_n^p = \left[\lim_{n\to\infty} a_n\right]^p$ provided $a_n \geq 0$ provided

These properties can be proved using Theorem above and the function limit properties we saw in Calculus I or we can prove them directly using the precise definition of a limit using nearly identical proofs of the function limit properties. Next, just as we had a Squeeze Theorem for function limits we also have one for sequences and it is pretty much identical to the function limit version.

Squeeze Theorem for Sequences

If $a_n \leq c_n \leq b_n$ for all $n > N$ for some N and $\lim_{n\to\infty} a_a$ $\lim_{n\to\infty}$ $b_n = L$ then $\lim_{n\to\infty} c_n = L$.

Note that in this theorem the "for all $n > N$ for some N" is really just telling us that we need to have $a_n \le c_n \le b_n$ for all sufficiently large n, but if it isn't true for the first few n that won't invalidate the theorem. As we'll see not all sequences can be written as functions that we can actually take the limit of. This will be especially true for sequences that alternate in signs. While we can always write these sequence terms as a function we simply don't know how to take the limit of a function like that. The following theorem will help with some of these sequences.

Theorem: If $\lim_{n\to\infty} |a_n| = 0$ then $\lim_{n\to\infty} a_n = 0$.

Note that in order for this theorem to hold the limit MUST be zero and it won't work for a sequence whose limit is not zero. This theorem is easy enough to prove so let's do that. Proof of Theorem: The main thing to this proof is to note that,

$$-|a_n| \le a_n \le |a_n|$$

Then note that,

$$\lim_{n\to\infty} (-|a_n|) = -\lim_{n\to\infty} |a_n| = 0$$

We then have $\lim_{n\to\infty} (-|a_n|) = \lim_{n\to\infty} |a_n| = 0$ and so by the Squeeze Theorem we must also have,

$$\lim_{n\to\infty} a_n = 0$$

The next theorem is a useful theorem giving the convergence/divergence and value (for when it's convergent) of a sequence that arises on occasion.

Theorem: The sequence $\left\{r^n\right\}_{n=0}^{\infty}$ converges if $-1 < r \le 1$ and diverges for all other value of r. Also,

$$\lim_{n\to\infty} r^n = \begin{cases} 0 & \text{if } -1 < r < 1 \\ 1 & \text{if } r = 1 \end{cases}$$

Here is a simple partial proof of this theorem.

Partial Proof of Theorem:

We'll do this by a series of cases although the last case will not be completely proven.

Case 1: r > 1

We know from Calculus I that $\lim\limits_{n\to\infty} r^x = \infty$ if $r > 1$ and so by Theorem above we also know that $\lim\limits_{n\to\infty} r^n = \infty$ and so the sequence diverges if $r > 1$.

Case 2: r = 1

In this case we have,

$$\lim_{n\to\infty} r^n = \lim_{n\to\infty} 1^n = \lim_{n\to\infty} 1 = 1$$

So, the sequence converges for $r = 1$ and in this case its limit is 1.

Case 3: 0 < r < 1

We know from Calculus I that $\lim\limits_{n\to\infty} r^x = 0$ if $0 < r < 1$ and so by Theorem above we also know that $\lim\limits_{n\to\infty} r^n = 0$ and so the sequence converges if $0 < r < 1$ and in this case its limit is zero.

Case 4: r = 0

In this case we have,

$$\lim_{n\to\infty} r^n = \lim_{n\to\infty} 0^n = \lim_{n\to\infty} 0 = 0$$

So, the sequence converges for $r = 0$ and in this case its limit is zero.

Case 5: – 1 < r < 0

First let's note that if $-1 < r < 0$ then $0 < |r| < 1$ then by Case 3 above we have,

$$\lim_{n\to\infty} |r^n| = \lim_{n\to\infty} |r^n| = 0$$

Theorem above now tells us that we must also have, $\lim_{n\to\infty} r^n = 0$ and so if $-1 < r < 0$ the sequence converges and has a limit of 0.

Case 6:

In this case the sequence is,

$$\{r^n\}_{n=0}^{\infty} = \{(-1)^n\}_{n=0}^{\infty}$$

$$= \{1,-1,1,-1,1-1,1-1,...\}_{n=0}^{\infty}$$

and hopefully it is clear that $\lim_{n\to\infty} (-1)^n$ doesn't exist. Recall that in order of this limit to exist the terms must be approaching a single value as n increases. In this case however the terms just alternate between 1 and –1 and so the limit does not exist.

So, the sequence diverges for $r = -1$.

Case 7: r < – 1

In this case we're not going to go through a complete proof. Let's just see what happens if we let $r = -2$ for instance. If we do that the sequence becomes,

$$\{r^n\}_{n=0}^{\infty} = \{(-2)^n\}_{n=0}^{\infty} = \{1, -2, 4, -8, 16, -32, ...\}_{n=0}^{\infty}$$

So, if $r = -2$ we get a sequence of terms whose values alternate in sign and get larger and larger and so $\lim_{n\to\infty} (-2)^n$ doesn't exist. It does not settle down to a single value as n increases nor do the terms ALL approach infinity. So, the sequence diverges for $r = -2$. We could do something similar for any value of r such that $r < -1$ and so the sequence diverges for $r < -1$.

Let's take a look at a couple of examples of limits of sequences.

Example: Determine if the following sequences converge or diverge. If the sequence converges determine its limit.

1. $\left\{\dfrac{3n^2-1}{10n+5n^2}\right\}_{n=2}^{\infty}$

2. $\left\{\dfrac{e^{2n}}{n}\right\}_{n=1}^{\infty}$

3. $\left\{\dfrac{(-1)^n}{n}\right\}_{n=1}^{\infty}$

4. $\left\{(-1)^n\right\}_{n=1}^{\infty}$

Solution:

1. $\left\{\dfrac{3n^2-1}{10n+5n^2}\right\}_{n=2}^{\infty}$

In this case all we need to do is recall the method that was developed in Calculus I to deal with the limits of rational functions

To do a limit in this form all we need to do is factor from the numerator and denominator the largest power of n, cancel and then take the limit.

$$\lim_{n\to\infty}\frac{3n^2-1}{10n+5n^2}=\lim_{n\to\infty}\frac{n^2\left(3-\frac{1}{n^2}\right)}{n^2\left(\frac{10}{n}+5\right)}=\lim_{n\to\infty}\frac{3-\frac{1}{n^2}}{\frac{10}{n}+5}=\frac{3}{5}$$

So the sequence converges and its limit is $\frac{3}{5}$.

2. $\left\{\dfrac{e^{2n}}{n}\right\}_{n=1}^{\infty}$

We will need to be careful with this one. We will need to use L'Hospital's Rule on this sequence. The problem is that L'Hospital's Rule only works on functions and not on sequences.

Normally this would be a problem, but we've got Theorem from above to help us out. Let's define

$$f(x) = \frac{e^{2x}}{x}$$

and note that,

$$f(n) = \frac{e^{2n}}{n}$$

Theorem says that all we need to do is take the limit of the function.

$$\lim_{n\to\infty} \frac{e^{2n}}{n} = \lim_{n\to\infty} \frac{e^{2x}}{x} = \lim_{n\to\infty} \frac{2e^{2x}}{1} = \infty$$

So, the sequence in this part diverges (to ∞).

More often than not we just do L'Hospital's Rule on the sequence terms without first converting to *x's* since the work will be identical regardless of whether we use x or n.

However, we really should remember that technically we can't do the derivatives while dealing with sequence terms.

3. $\left\{ \frac{(-1)^n}{n} \right\}_{n=1}^{\infty}$

We will also need to be careful with this sequence. We might be tempted to just say that the limit of the sequence terms is zero (and we'd be correct). However, technically we can't take the limit of sequences whose terms alternate in sign, because we don't know how to do limits of functions that exhibit that same behavior.

Also, we want to be very careful to not rely too much on intuition with these problems. So, let's work this one by the book. We will need to use Theorem on this problem. To this we'll first need to compute,

$$\lim_{n\to\infty} \left| \frac{(-1)^n}{n} \right| = \lim_{n\to\infty} \frac{1}{n} = 0$$

Therefore, since the limit of the sequence terms with absolute value bars on them goes to zero we know by Theorem that,

$$\lim_{n\to\infty} \frac{(-1)^n}{n} = 0$$

Which also means that the sequence converges to a value of zero.

4. $\left\{(-1)^n\right\}_{n=0}^{\infty}$

For this theorem note that all we need to do is realize that this is the sequence in Theorem above using $r = -1$. So, by Theorem this sequence diverges. We now need to give a warning about misusing Theorem. Theorem only works if the limit is zero. If the limit of the absolute value of the sequence terms is not zero then the theorem will not hold. The last part of the previous example is a good example of this (and in fact this warning the whole reason that part is there). Notice that

$$\lim_{n\to\infty} \left|(-1)^n\right| = \lim_{n\to\infty} 1 = 1$$

and yet, $\lim_{n\to\infty} (-1)^n$ doesn't even exist let alone equal 1.So, be careful using this Theorem. Before moving onto the next section we need to give one more theorem that we'll need for a proof down the road. Theorem: For the sequence $\{a_n\}$ if both $\lim_{n\to\infty} a_{2n} = L$ and $\lim_{n\to\infty} a_{2n+1} = L$ then $\{a_n\}$ is convergent and $\lim_{n\to\infty} a_n = L$.

Proof of Theorem

Let å > 0.

Then since $\lim_{n\to\infty} a_{2n} = L$ there is an $N_1 > 0$ such that if $n > N_1$ we know that,

$$|a2n - L| < å$$

Likewise, because,

$$\lim_{n\to\infty} a_{2n+1} = L$$

there is an $N_2 > 0$ such that if $n > N_2$ we know that,

$$| a_{2n+1} - L | < å$$

Now, let $N = \max\{2N_1, 2N_2 + 1\}$ and let $n > N$. Then either $an = a_{2k}$ for some $k > N_1$ or $a_n = a_{2k+1}$ for some $k > N_2$ and so in either case we have that,

$$| a_n - L | \quad < å$$

Therefore, $\lim_{n\to\infty} a_n = L$ and so $\{a_n\}$ is convergent.

More on Sequences

In the previous section we introduced the concept of a sequence and talked about limits of sequences and the idea of convergence and divergence for a sequence. In this section we want to take a quick look at some ideas involving sequences.

Let's start off with some terminology and definitions.

Given any sequence $\{a_n\}$ we have the following.

1. We call the sequence increasing if $a_n < a_{n+1}$ for every n.
2. We call the sequence decreasing if $a_n > a_{n+1}$ for every n.
3. If $\{a_n\}$ is an increasing sequence or $\{a_n\}$ is a decreasing sequence we call it monotonic.
4. If there exists a number m such that $m \leq a_n$ for every n we say the sequence is bounded below. The number m is sometimes called a lower bound for the sequence.
5. If there exists a number M such that $a_n \leq M$ for every n we say the sequence is bounded above. The number M is sometimes called an upper bound for the sequence.
6. If the sequence is both bounded below and bounded above we call the sequence bounded.

Note that in order for a sequence to be increasing or decreasing it must be increasing/decreasing for every n. In other words, a sequence that increases for three terms and then decreases for the rest of the terms is NOT a decreasing sequence!

Also note that a monotonic sequence must always increase or it must always decrease. Before moving on we should make a quick point about the bounds for a sequence that is bounded above and/or below. We'll make the point about lower bounds, but we could just as easily make it about upper bounds. A sequence is bounded below if we can find any number m such that $m \leq a_n$ for every n. Note however that if we find one number m to use for a lower bound then any number smaller than m will also be a lower bound.

Also, just because we find one lower bound that doesn't mean there won't be a "better" lower bound for the sequence than the one we found. In other words, there are an infinite number of lower bounds for a sequence that is bounded below, some will be better than others.

In my class all that I'm after will be a lower bound. I don't necessarily need the best lower bound, just a number that will be a lower bound for the sequence.

Chapter 7

Pre-calculus Integration

The first documented systematic technique capable of determining integrals is the method of exhaustion of the ancient Greek astronomer Eudoxus (*ca.* 370 BC), which sought to find areas and volumes by breaking them up into an infinite number of shapes for which the area or volume was known.

This method was further developed and employed by Archimedes in the 3rd century BC and used to calculate areas for parabolas and an approximation to the area of a circle. Similar methods were independently developed in China around the 3rd century AD by Liu Hui, who used it to find the area of the circle. This method was later used in the 5th century by Chinese father-and-son mathematicians Zu Chongzhi and Zu Geng to find the volume of a sphere (Shea 2007; Katz 2004, pp. 125–126).

The next significant advances in integral calculus did not begin to appear until the 16th century. At this time the work of Cavalieri with his *method of indivisibles*, and work by Fermat, began to lay the foundations of modern calculus, with Cavalieri computing the integrals of x^n up to degree $n - 9$ in Cavalieri's quadrature formula.

Further steps were made in the early 17th century by Barrow and Torricelli, who provided the first hints of a connection between integration and differentiation. Barrow provided the first proof of the fundamental theorem of calculus. Wallis generalized Cavalieri's method, computing integrals of x to a general power, including negative powers and fractional powers.

Newton and Leibniz

The major advance in integration came in the 17th century with the independent discovery of the fundamental theorem of calculus by Newton and Leibniz. The theorem demonstrates a connection between integration and differentiation. This connection, combined with the comparative ease of differentiation, can be exploited to calculate integrals. In particular, the fundamental theorem of calculus allows one to solve a much broader class of problems. Equal in importance is the comprehensive mathematical framework that both Newton and Leibniz developed. Given the name infinitesimal calculus, it allowed for precise analysis of functions within continuous domains. This framework eventually became modern calculus, whose notation for integrals is drawn directly from the work of Leibniz.

Formalizing Integrals

While Newton and Leibniz provided a systematic approach to integration, their work lacked a degree of rigour. Bishop Berkeley memorably attacked the vanishing increments used by Newton, calling them "ghosts of departed quantities". Calculus acquired a firmer footing with the development of limits. Integration was first rigorously formalized, using limits, by Riemann.

Although all bounded piecewise continuous functions are Riemann integrable on a bounded interval, subsequently more general functions were considered – particularly in the context of Fourier analysis – to which Riemann's definition does not apply, and Lebesgue formulated a different definition of integral, founded in measure theory (a subfield of real analysis). Other definitions of integral, extending Riemann's and Lebesgue's approaches, were proposed. These approaches based on the real number system are the ones most common today, but alternative approaches exist, such as a definition of integral as the standard part of an infinite Riemann sum, based on the hyperreal number system.

Historical Notation

Isaac Newton used a small vertical bar above a variable to indicate integration, or placed the variable inside a box. The vertical bar was easily confused with $\dot{x}$ or x' , which Newton used to indicate differentiation, and the box notation was difficult for printers to reproduce, so these notations were not widely adopted.

The modern notation for the indefinite integral was introduced by Gottfried Leibniz in 1675 (Burton 1988, p. 359; Leibniz 1899, p. 154). He adapted the integral symbol, *ʃ*, from the letter *ʃ* (long s), standing for *summa* (written as *ʃumma*; Latin for "sum" or "total"). The modern notation for the definite integral, with limits above and below the integral sign, was first used by Joseph Fourier in *Mémoires* of the French Academy around 1819–20, reprinted in his book of 1822.

Terminology and Notation

The simplest case, the integral over *x* of a real-valued function *f*(*x*), is written as

$$\int f(x)dx.$$

The integral sign *ʃ* represents integration. The *dx* indicates that we are integrating over *x*; *dx* is called the variable of integration. In correct mathematical typography, the *dx* is separated from the integrand by a space (as shown). Some authors use an upright *d* (that is, d*x* instead of *dx*). Inside the *ʃ...dx* is the expression to be integrated, called the *integrand.* In this case the integrand is the function *f*(*x*). Because there is no domain specified, the integral is called an *indefinite integral.*

When integrating over a specified domain, we speak of a *definite integral.* Integrating over a domain *D* is written as

$\int_D f(x)dx$, or $\int_a^b f(x)dx$ if the domain is an interval [*a*, *b*] of *x*;

The domain *D* or the interval [*a*, *b*] is called the *domain of integration.*

If a function has an integral, it is said to be *integrable.* In general, the integrand may be a function of more than one variable, and the domain of integration may be an area, volume, a higher dimensional region, or even an abstract space that does not have a geometric structure in any usual sense (such as a sample space in probability theory). In the modern Arabic mathematical notation, which aims at pre-university levels of education in the Arab world and is written from right to left, a reflected integral symbol ∫ is used (W3C 2006).

The variable of integration dx has different interpretations depending on the theory being used. It can be seen as strictly a notation indicating that x is a dummy variable of integration; if the integral is seen as a Riemann sum, dx is a reflection of the weights or widths d of the intervals of x; in Lebesgue integration and its extensions, dx is a measure; in non-standard analysis, it is an infinitesimal; or it can be seen as an independent mathematical quantity, a differential form. More complicated cases may vary the notation slightly. In Leibniz's notation, dx is interpreted an infinitesimal change in x, but his interpretation lacks rigour in the end. Nonetheless Leibniz's notation is the most common one today; and as few people are in need of full rigour, even his interpretation is still used in many settings.

Integrals appear in many practical situations. If a swimming pool is rectangular with a flat bottom, then from its length, width, and depth we can easily determine the volume of water it can contain (to fill it), the area of its surface (to cover it), and the length of its edge (to rope it). But if it is oval with a rounded bottom, all of these quantities call for integrals. Practical approximations may suffice for such trivial examples, but precision engineering (of any discipline) requires exact and rigorous values for these elements.

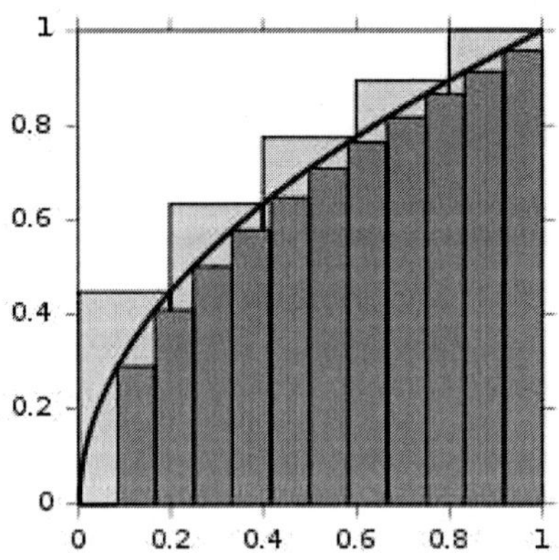

Approximations to integral of √x from 0 to 1, with % 5 right samples (above) and % 12 left samples (below)

To start off, consider the curve $y = f(x)$ between $x = 0$ and $x = 1$ with $f(x) = \sqrt{x}$. We ask:

What is the area under the function f, in the interval from 0 to 1?

and call this (yet unknown) area the integral of f. The notation for this integral will be

$$\int_0^1 \sqrt{x}\,dx.$$

As a first approximation, look at the unit square given by the sides $x = 0$ to $x = 1$ and $y = f(0) = 0$ and $y = f(1) = 1$. Its area is exactly 1. As it is, the true value of the integral must be somewhat less. Decreasing the width of the approximation rectangles shall give a better result; so cross the interval in five steps, using the approximation points 0, 1/5, 2/5, and so on to 1. Fit a box for each step using the right end height of each curve piece, thus √(1D†5), √(2D†5), and so on to √1 = 1. Summing the areas of these rectangles, we get a better approximation for the sought integral, namely

$$\sqrt{\frac{1}{5}}\left(\frac{1}{5}-0\right)+\sqrt{\frac{2}{5}}\left(\frac{2}{5}-\frac{1}{5}\right)+\cdots+\sqrt{\frac{5}{5}}\left(\frac{5}{5}-\frac{4}{5}\right)\approx 0.7497.$$

Notice that we are taking a sum of finitely many function values of f, multiplied with the differences of two subsequent approximation points. We can easily see that the approximation is still too large.

Using more steps produces a closer approximation, but will never be exact: replacing the 5 subintervals by twelve as depicted, we will get an approximate value for the area of 0.6203, which is too small. The key idea is the transition from adding *finitely many* differences of approximation points multiplied by their respective function values to using infinitely many fine, or *infinitesimal* steps.

As for the *actual calculation of integrals*, the fundamental theorem of calculus, due to Newton and Leibniz, is the fundamental link between the operations of differentiating and integrating. Applied to the square root curve, $f(x) = x^{1/2}$, it says to look at the antiderivative $F(x) = (2/3)x^{3/2}$, and simply take $F(1) - F(0)$, where 0 and 1 are the boundaries of the interval [0,1]. So the *exact* value of the area under the curve is computed formally as

$$\int_0^1 \sqrt{x}\,dx = \int_0^1 x^{1/2}\,dx = F(1) - F(0) = 2/3.$$

(This is a case of a general rule, that for $f(x) = x^q$, with $q \neq -1$, the related function, the so-called antiderivative is $F(x) = x^{q+1}/(q + 1)$.)

The notation

$$\int f(x)dx$$

conceives the integral as a weighted sum, denoted by the elongated *s*, of function values, $f(x)$, multiplied by infinitesimal step widths, the so-called *differentials*, denoted by dx. The multiplication sign is usually omitted.

Historically, after the failure of early efforts to rigorously interpret infinitesimals, Riemann formally defined integrals as a limit of weighted sums, so that the dx suggested the limit of a difference (namely, the interval width). Shortcomings of Riemann's dependence on intervals and continuity motivated newer definitions, especially the Lebesgue integral, which is founded on an ability to extend the idea of "measure" in much more flexible ways. Thus the notation

$$\int_A f(x)d\mu$$

refers to a weighted sum in which the function values are partitioned, with μ measuring the weight to be assigned to each value. Here A denotes the region of integration.

Differential geometry, with its "calculus on manifolds", gives the familiar notation yet another interpretation. Now $f(x)$ and dx become a differential form, $\omega = f(x)\,dx$, a new differential operator d, known as the exterior derivative is introduced, and the fundamental theorem becomes the more general Stokes' theorem,

$$\int_A d\omega = \int_{\partial A} \omega,$$

from which Green's theorem, the divergence theorem, and the fundamental theorem of calculus follow. More recently, infinitesimals have reappeared with rigor, through modern innovations such as non-standard analysis. Not only do these methods vindicate the intuitions of the pioneers; they also lead to new mathematics. Although there are differences between these conceptions of integral, there is considerable overlap. Thus, the area of the surface of the oval swimming pool can be handled as a geometric ellipse, a sum of infinitesimals, a Riemann integral, a Lebesgue integral, or as a manifold with a differential form. The calculated result will be the same for all.

Formal Definitions

There are many ways of formally defining an integral, not all of which are equivalent. The differences exist mostly to deal with differing special cases which may not be integrable under other definitions, but also occasionally for pedagogical reasons. The most commonly used definitions of integral are Riemann integrals and Lebesgue integrals.

Other Integrals

Riemann–Stieltjes Integral

In mathematics, the Riemann–Stieltjes integral is a generalization of the Riemann integral, named after Bernhard Riemann and Thomas Joannes Stieltjes.

Definition

The Riemann–Stieltjes integral of a real-valued function f of a real variable with respect to a real function g is denoted by

$$\int_a^b f(x)dg(x)$$

and defined to be the limit, as the mesh of the partition

$$P = \{a = x_0 < x_1 < \cdots < x_n = b\}$$

of the interval $[a, b]$ approaches zero, of the approximating sum

$$S(P, f, g) = \sum_{i=0}^{n-1} f(c_i)(g(x_{i+1}) - g(x_i))$$

where c_i is in the i-th subinterval $[x_i, x_{i+1}]$. The two functions f and g are respectively called the integrand and the integrator.

The "limit" is here understood to be a number A (the value of the Riemann–Stieltjes integral) such that for every $\varepsilon > 0$, there exists $\delta > 0$ such that for every partition P with mesh$(P) < \delta$, and for every choice of points c_i in $[x_i, x_{i+1}]$,

$$|S(P, f, g) - A| < \varepsilon.$$

Generalized Riemann–Stieltjes Integral

A slight generalization, introduced by Pollard (1920) and now standard in analysis, is to consider in the above definition partitions P that *refine* another partition P_ε, meaning that P arises from P_ε by the addition of points, rather than from partitions with a finer mesh.

Specifically, the generalized Riemann–Stieltjes integral of f with respect to g is a number A such that for every $\varepsilon > 0$ there exists a partition P_ε such that for every partition P that refines P_ε,

$$|S(P, f, g) - A| < \varepsilon$$

for every choice of points c_i in $[x_i, x_{i+1}]$.

This generalization exhibits the Riemann–Stieltjes integral as the Moore–Smith limit on the directed set of partitions of $[a, b]$ (McShane 1952). Hildebrandt (1938) calls it the Pollard–Moore–Stieltjes integral.

Darboux Sums

The Riemann–Stieltjes integral can be efficiently handled using an appropriate generalization of Darboux sums. For a partition P and a nondecreasing function g on $[a, b]$ define the upper Darboux sum of f with respect to g by

$$U(P, f, g) = \sum_{i=1}^{n} \sup_{x \in [x_i, x_{i+1}]} f(x)(g(x_{i+1}) - g(x_i))$$

and the lower sum by

$$L(P, f, g) = \sum_{i=1}^{n} \inf_{x \in [x_i, x_{i+1}]} f(x)(g(x_{i+1}) - g(x_i)).$$

Then the generalized Riemann–Stieltjes of f with respect to g exists if and only if, for every $\varepsilon > 0$, there exists a partition P such that

$$U(P, f, g) - L(P, f, g) < \varepsilon.$$

Furthermore, f is Riemann–Stieltjes integrable with respect to g (in the classical sense) if

$$\lim_{\text{mesh}(P)\to 0} [U(P, f, g) - L(P, f, g)] = 0.$$

Properties and Relation to the Riemann Integral

If g should happen to be everywhere differentiable, then the Riemann–Stieltjes integral may still be different from the Riemann integral of $f(x)g'(x)$ given by

$$\int_a^b f(x)g'(x)dx,$$

for example, if the derivative is unbounded. But if the derivative is continuous, they will be the same. This condition is also satisfied if g is the (Lebesgue) integral of its derivative; in this case g is said to be absolutely continuous.

However, g may have jump discontinuities, or may have derivative zero *almost* everywhere while still being continuous and increasing (for example, g could be the Cantor function), in either of which cases the Riemann–Stieltjes integral is not captured by any expression involving derivatives of g.

The Riemann–Stieltjes integral admits integration by parts in the form

$$\int_a^b f(x)dg(x) = f(b)g(b) - f(a)g(a) - \int_a^b g(x)df(x).$$

and the existence of either integral implies the existence of the other (Hille & Phillips 1974, §3.3).

Existence of the Integral

The best simple existence theorem states that if f is continuous and g is of bounded variation on $[a, b]$, then the integral exists. A function g is of bounded variation if and only if it is the difference between two monotone functions. If g is not of bounded variation, then there will be continuous functions which cannot be integrated with respect to g. In general, the integral is not well-defined if f and g share any points of discontinuity, but this sufficient condition is not necessary. On the other hand, a classical result of Young (1936) states that the integral is well-defined if f is α-Hölder continuous and g is β-Hölder continuous with $\alpha + \beta > 1$.

Application to Probability Theory

If g is the cumulative probability distribution function of a random variable X that has a probability density function with respect to Lebesgue measure, and f is any function for which the expected value $E(|f(X)|)$ is finite, then the probability density function of X is the derivative of g and we have

$$E(f(X)) = \int_{-\infty}^{\infty} f(x)g'(x)dx.$$

But this formula does not work if X does not have a probability density function with respect to Lebesgue measure. In particular, it does not work if the distribution of X is discrete (i.e., all of the probability is accounted for by point-masses), and even if the cumulative distribution function g is continuous, it does not work if g fails to be absolutely continuous (again, the Cantor function may serve as an example of this failure). But the identity

$$E(f(X)) = \int_{-\infty}^{\infty} f(x)dg(x)$$

holds if g is *any* cumulative probability distribution function on the real line, no matter how ill-behaved. In particular, no matter how ill-behaved the cumulative distribution function g of a random variable X, if the moment $E(X^n)$ exists, then it is equal to

$$E(X^n) = \int_{-\infty}^{\infty} x^n dg(x).$$

Application to Functional Analysis

The Riemann–Stieltjes integral appears in the original formulation of F. Riesz's theorem which represents the dual space of the Banach space $C[a,b]$ of continuous functions in an interval $[a,b]$ as Riemann–Stieltjes integrals against functions of bounded variation. Later, that theorem was reformulated in terms of measures. The Riemann–Stieltjes integral also appears in the formulation of the spectral theorem for (non-compact) self-adjoint (or more generally, normal) operators in a Hilbert space. In this theorem, the integral is considered with respect to a spectral family of projections.

Generalization

An important generalization is the Lebesgue–Stieltjes integral which generalizes the Riemann–Stieltjes integral in a way analogous to how the Lebesgue integral generalizes the Riemann integral. If improper Riemann–Stieltjes integrals are allowed, the Lebesgue integral is not strictly more general than the Riemann–Stieltjes integral.

The Riemann–Stieltjes integral also generalizes to the case when either the integrand f or the integrator g take values in a Banach space. If $g : [a,b] \to X$ takes values in the Banach space X, then it is natural to assume that it is of strongly bounded variation, meaning that

$$\sup \sum_i \| g(t_i) - g(t_{i+1}) \|_X < \infty$$

the supremum being taken over all finite partitions

$$a = t_0 \leq t_1 \leq \cdots \leq t_n = b$$

of the interval $[a,b]$. This generalization plays a role in the study of semigroups, via the Laplace–Stieltjes transform.

Lebesgue–Stieltjes Integration

In measure-theoretic analysis and related branches of mathematics, Lebesgue–Stieltjes integration generalizes Riemann–Stieltjes and Lebesgue integration, preserving the many advantages of the former in a more general measure-theoretic framework. The Lebesgue–Stieltjes integral is the ordinary Lebesgue integral with respect to a measure known as the Lebesgue–Stieltjes measure, which may be associated to any function of bounded variation on the real line. The Lebesgue–Stieltjes measure is a regular Borel measure, and conversely every regular Borel measure on the real line is of this kind. Lebesgue–Stieltjes integrals, named for Henri Leon Lebesgue and Thomas Joannes Stieltjes, are also known as Lebesgue–Radon integrals or just Radon integrals, after Johann Radon, to whom much of the theory is due. They find common application in probability and stochastic processes, and in certain branches of analysis including potential theory.

Definition

The Lebesgue–Stieltjes integral

$$\int_a^b f(x)dg(x)$$

is defined when $f : [a,b] \to \mathrm{R}$ is Borel-measurable and bounded and $g : [a,b] \to \mathrm{R}$ is of bounded variation in $[a,b]$ and right-continuous, or when f is non-negative and g is monotone and right-continuous. To start, assume that f is non-negative and g is monotone non-decreasing and right-continuous. Define $w((s,t]) := g(t) - g(s)$ and $w(\{a\}) := 0$ (Alternatively, the construction works for g left-continuous, $w([s,t)) := g(t) - g(s)$ and $w(\{b\}) := 0$).

By Carathéodory's extension theorem, there is a unique Borel measure μ_g on $[a,b]$ which agrees with w on every interval I. The measure μ_g arises from an outer measure (in fact, a metric outer measure) given by

$$\mu_g(E) = \inf\{\sum_i \mu_g(I_i) \mid E \subset \bigcup_i I_i\}$$

the infimum taken over all coverings of E by countably many semiopen intervals. This measure is sometimes called the Lebesgue–Stieltjes measure associated with g.

The Lebesgue–Stieltjes integral

$$\int_a^b f(x)dg(x)$$

is defined as the Lebesgue integral of f with respect to the measure μ_g in the usual way. If g is non-increasing, then define

$$\int_a^b f(x)dg(x) := -\int_a^b f(x)d(-g)(x),$$

the latter integral being defined by the preceding construction.

If g is of bounded variation and f is bounded, then it is possible to write

$$g(x) = g_1(x) - g_2(x)$$

where $g_1(x) := Vx$

ag is the total variation of g in the interval $[a,x]$, and $g_2(x) = g_1(x) - g(x)$. Both g_1 and g_2 are monotone non-decreasing. Now the Lebesgue–Stieltjes integral with respect to g is defined by

$$\int_a^b f(x)dg(x) = \int_a^b f(x)dg_1(x) - \int_a^b f(x)dg_2(x),$$

where the latter two integrals are well-defined by the preceding construction.

Daniell Integral

An alternative approach (Hewitt & Stromberg 1965) is to define the Lebesgue–Stieltjes integral as the Daniell integral that extends the usual Riemann–Stieltjes integral. Let g be a non-increasing right-continuous function on $[a,b]$, and define $I(f)$ to be the Riemann–Stieltjes integral

$$I(f) = \int_a^b f(x)dg(x)$$

for all continuous functions f. The functional I defines a Radon measure on $[a,b]$. This functional can then be extended to the class of all non-negative functions by setting

$$\overline{I}(h) = \sup\{I(f) \mid f \in C[a,b], 0 \le f \le h\}$$

and

$$\overline{\overline{I}}(h) - \inf\{I(f) \mid f \in C[a,b], h < f\}.$$

For Borel measurable functions, one has

$$\overline{I}(h) = \overline{\overline{I}}(h),$$

and either side of the identity then defines the Lebesgue–Stieltjes integral of h. The outer measure μ_g is defined via

$$\mu_g(A) = \overline{\overline{I}}(\chi_A)$$

where χ_A is the indicator function of A.

Integrators of bounded variation are handled as above by decomposing into positive and negative variations.

Example

Suppose that $\gamma:[a,b]\to\mathbb{R}^2$ is a rectifiable curve in the plane and $\rho:\mathbb{R}^2\to[0,\infty)$ is Borel measurable. Then we may define the length of γ with respect to the Euclidean metric weighted by ρ to be $\int_a^b \rho(\gamma(t))d\ell(t)$, where $\ell(t)$ is the length of the restriction of γto $[a,t]$. This is sometimes called the ρ-length of γ. This notion is quite useful for various applications: for example, in muddy terrain the speed in which a person can move may depend on how deep the mud is. If $\rho(z)$ denotes the inverse of the walking speed at or near z, then the ρ-length of γis the time it would take to traverse γ. The concept of extremal length uses this notion of the ρ-length of curves and is useful in the study of conformal mappings.

Integration by Parts

A function f is said to be "regular" at a point if the right $f(a+)$ and left hand limits and $f(a-)$ exist, and the function takes the average value,

$$f(a)=\frac{1}{2}\big(f(a-)+f(a+)\big),$$

at the limiting point. Given two functions *U* and *V* of finite variation, if at each point either *U* or *V* is continuous, or if both *U* and *V* are regular, then there is an integration by parts formula for the Lebesgue–Stieltjes integral:

$$\int_a^b U\,dV+\int_a^b V\,dU=U(b+)V(b+)-U(a-)V(a-),$$

where $b>a$. Under a slight generalization of this formula, the extra conditions on *U* and *V* can be dropped.

An alternative result, of significant importance in the theory of Stochastic calculus is the following. Given two functions *U* and V of finite variation, which are both right-continuous and have left-limits (they are cadlag functions) then

$$U(t)V(t) = U(0)V(0) + \int_{(0,t]} U\,dV + \int_{(0,t]} V\,dU + \sum_{u\in(0,t]} \Delta U_u \Delta V_u,$$

where $\Delta U_u = U(t) - U(t-)$. This result can be seen as a precursor to Ito's lemma, and is of use in the general theory of Stochastic integration. The final term is $\Delta U(t)\Delta V(t) = d[U,V]$, which arises from the quadratic covariation of U and V.

Related Concepts

Lebesgue Integration: When $g(x) = x$ for all real x, then μ_g is the Lebesgue measure, and the Lebesgue–Stieltjes integral of f with respect to g is equivalent to the Lebesgue integral of f.

Riemann–Stieltjes Integration and Probability Theory

Where f is a continuous real-valued function of a real variable and v is a non-decreasing real function, the Lebesgue–Stieltjes integral is equivalent to the Riemann–Stieltjes integral, in which case we often write

$$\int_a^b f(x)dv(x)$$

for the Lebesgue–Stieltjes integral, letting the measure μ_v remain implicit. This is particularly common in probability theory when v is the cumulative distribution function of a real-valued random variable X, in which case

$$\int_{\infty}^{\infty} f(x)dv(x) = \mathrm{E}[f(X)].$$

Daniell Integral

In mathematics, the Daniell integral is a type of integration that generalizes the concept of more elementary versions such as the Riemann integral to which students are typically first introduced.

One of the main difficulties with the traditional formulation of the Lebesgue integral is that it requires the initial development of a workable measure theory before any useful results for the integral can be obtained. However, an alternative approach is available, developed by Percy J. Daniell (1918) that does not suffer from this deficiency, and has a few significant advantages over the traditional formulation, especially as the integral is generalized into higher dimensional spaces and further generalizations such as the Stieltjes integral. The basic idea involves the axiomatization of the integral.

The Daniell Axioms

We start by choosing a family *H* of bounded real functions (called *elementary functions*) defined over some set *X*, that satisfies these two axioms:

1. *H* is a linear space with the usual operations of addition and scalar multiplication.
2. If a function $h(x)$ is in *H*, so is its absolute value $|h(x)|$.

In addition, every function *h* in *H* is assigned a real number Ih, which is called the *elementary integral* of *h*, satisfying these three axioms:

1. Linearity. If *h* and *k* are both in H, αand βand are any two real numbers, then $I(\alpha h+\beta k)=\alpha Ih+\beta Ik$.
2. Nonnegativity. If $h(x)\geq 0$, then $Ih\geq 0$.
3. Continuity. If $h_n(x)$ is a nonincreasing sequence (i.e. $h_1\geq\cdots\geq h_k\geq\cdots$) of functions in H that converges to 0 for all *x* in X, then $Ih_n\to 0$.

That is, we define a continuous non-negative linear functional over the space of elementary functions.

These elementary functions and their elementary integrals may be any set of functions and definitions of integrals over these functions which satisfy these axioms.

The family of all step functions evidently satisfies the above axioms for elementary functions. Defining the elementary integral of the family of step functions as the (signed) area underneath a step function evidently satisfies the given axioms for an elementary integral. Applying the construction of the Daniell integral described further below using step functions as elementary functions produces a definition of an integral equivalent to the Lebesgue integral.

Using the family of all continuous functions as the elementary functions and the traditional Riemann integral as the elementary integral is also possible, however, this will yield an integral that is also equivalent to Lebesgue's definition.

Doing the same, but using the Riemann–Stieltjes integral, along with an appropriate function of bounded variation, gives a definition of integral equivalent to the Lebesgue–Stieltjes integral.

Sets of measure zero may be defined in terms of elementary functions as follows. A set Z which is a subset of X is a set of measure zero if for any $\epsilon > 0$, there exists a nondecreasing sequence of nonnegative elementary functions $h_p(x)$ in H such that $Ih_p < \epsilon$ and $\sup_p h_p(x) \geq 1$ on Z.

A set is called a set of full measure if its complement, relative to X, is a set of measure zero. We say that if some property holds at every point of a set of full measure (or equivalently everywhere except on a set of measure zero), it holds almost everywhere.

Definition of the Daniell Integral

We can then proceed to define a larger class of functions, based on our chosen elementary functions, the class L^+, which is the family of all functions that are the limit of a nondecreasing sequence h_n of elementary functions almost everywhere, such that the set of integrals Ih_n is bounded. The integral of a function f in L^+ is defined as:

$$If = \lim_{n \to \infty} Ih_n$$

It can be shown that this definition of the integral is well-defined, i.e. it does not depend on the choice of sequence h_n.

However, the class L^+ is in general not closed under subtraction and scalar multiplication by negative numbers, but we can further extend it by defining a wider class of functions L such that every function $\phi(x)$ can be represented on a set of full measure as the difference $\phi = f - g$, for some functions f and g in the class L^+. Then the integral of a function $\phi(x)$ can be defined as:

$$\int_X \phi(x)dx = If - Ig$$

Again, it may be shown that this integral is well-defined, i.e. it does not depend on the decomposition of into f and g. This is the final construction of the Daniell integral.

Properties

Nearly all of the important theorems in the traditional theory of the Lebesgue integral, such as Lebesgue's dominated convergence theorem, the Riesz–Fischer theorem, Fatou's lemma, and Fubini's theorem may also readily be proved using this construction. Its properties are identical to the traditional Lebesgue integral. Measures from the Daniell integral: Because of the natural correspondence between sets and functions, it is also possible to use the Daniell integral to construct a measure theory. If we take the characteristic function $\chi(x)$ of some set, then its integral may be taken as the measure of the set. This definition of measure based on the Daniell integral can be shown to be equivalent to the traditional Lebesgue measure.

Advantages Over the Traditional Formulation

This method of constructing the general integral has a few advantages over the traditional method of Lebesgue, particularly in the field of functional analysis. The Lebesgue and Daniell constructions are equivalent, as pointed out above, if ordinary finite-valued step functions are chosen as elementary functions. However, as one tries to extend the definition of the integral into more complex domains (e.g. attempting to define the integral of a linear functional), one runs into practical difficulties using Lebesgue's construction that are alleviated with the Daniell approach.

The Polish mathematician Jan Mikusinski has made an alternative and more natural formulation of Daniell integration by using the notion of absolutely convergent series. His formulation works for Bochner integral (Lebesgue integral for mappings taking values in Banach spaces). Mikusinski's lemma allows one to define integral without mentioning null sets.

He also proved change of variables theorem for multiple integral for Bochner integrals and Fubini's theorem for Bochner integrals using Daniell integration. The book by Asplund and Bungart carries a lucid treatment of this approach for real valued functions. It also offers a proof of an abstract Radon–Nikodym theorem using Daniell–Mikusinski approach.

Haar Integral

Using the general theory of Lebesgue integration, one can then define an integral for all Borel measurable functions *f* on *G*. This integral is called the Haar integral. If μ is a left Haar measure, then

$$\int_G f(sx)\, d\mu(x) = \int_G f(x)\, d\mu(x)$$

for any integrable function *f*. This is immediate for characteristic functions, being essentially the definition of left invariance.

Examples

- A Haar measure on the topological group (R, +) which takes the value 1 on the interval [0,1] is equal to the restriction of Lebesgue measure to the Borel subsets of R. This can be generalized to (R^n, +).
- If *G* is the group of nonzero real numbers with multiplication as operation, then a Haar measure μ is given by

$$\mu(S) = \int_S \frac{1}{|t|} dt$$

for any Borel subset *S* of the nonzero reals.

This generalizes to the following:

For *G* = *GL*(n,R), any left Haar measure is a right Haar measure and one such measure μ is given by

$$\mu(S)=\int_S \frac{1}{|\det(X)|^n}dX$$

where dX denotes the Lebesgue measure on $\mathbb{R}n^2$, the set of all $n \times n$-matrices. This follows from the change of variables formula.

More generally, on any Lie group of dimension d a left Haar measure can be associated with any non-zero left-invariant d-form ω, as the *Lebesgue measure* $|\omega|$; and similarly for right Haar measures.

This means also that the modular function can be computed, as the absolute value of the determinant of the adjoint representation.

In order to define a Haar measure μ on the unit circle T, consider the function f from $[0,2\pi]$ onto T defined by $f(t) = (\cos(t),\sin(t))$. Then μ can be defined by

$$\mu(S)=\frac{1}{2\pi}m\left(f^{-1}(S)\right),$$

where m is the Lebesgue measure. The factor $(2\pi)^{-1}$ is chosen so that $\mu(T) = 1$.

If G is the group of non-null quaternions, then G can be seen as an open subset of R^4. A Haar measure μ is given by

$$\mu(S)=\int_S \frac{1}{(x^2+y^2+z^2+w^2)^2}dx\,dy\,dz\,dw$$

where $dx\ dy\ dz\ dw$ denotes the Lebesgue measure in R^4 and S is a Borel subset of G.

Henstock–Kurzweil Integral

In mathematics, the Henstock–Kurzweil integral, also known as the Denjoy integral (pronounced [dQÞÈ'wa]) and the Perron integral, is one of a number of definitions of the integral of a function. It is a generalization of the Riemann integral which in some situations is more useful than the Lebesgue integral. This integral was first defined by Arnaud Denjoy (1912). Denjoy was interested in a definition that would allow one to integrate functions like

$$f(x) = \frac{1}{x}\sin\left(\frac{1}{x^3}\right).$$

This function has a singularity at 0, and is not Lebesgue integrable. However, it seems natural to calculate its integral except over [-ε,δ] and then let ε, δ → 0.

Trying to create a general theory Denjoy used transfinite induction over the possible types of singularities which made the definition quite complicated. Other definitions were given by Nikolai Luzin (using variations on the notions of absolute continuity), and by Oskar Perron, who was interested in continuous major and minor functions. It took a while to understand that the Perron and Denjoy integrals are actually identical.

Later, in 1957, the Czech mathematician Jaroslav Kurzweil discovered a new definition of this integral elegantly similar in nature to Riemann's original definition which he named the gauge integral; the theory was developed by Ralph Henstock.

Due to these two important mathematicians, it is now commonly known as the Henstock–Kurzweil integral. The simplicity of Kurzweil's definition made some educators advocate that this integral should replace the Riemann integral in introductory calculus courses, but this idea has not gained traction.

Definition

Henstock's definition is as follows:

Given a tagged partition P of $[a, b]$, say

$$a = u_0 < u_1 < \cdots < u_n = b,\ t_i \in [u_{i-1}, u_i]$$

and a positive function

$$\delta : [a,b] \to (0,\infty),$$

which we call a *gauge*, we say P is δ-fine if

$$\forall i\ \ t_i - \delta(t_i) < u_{i-1} \le t_i \le u_i < t_i + \delta(t_i).$$

For a tagged partition P and a function

$$f : [a,b] \to \mathbb{R}$$

we define the Riemann sum to be

$$\sum_{P} f = \sum_{i=1}^{n} (u_i - u_{i-1}) f(t_i).$$

Given a function

$$f : [a,b] \to \mathbb{R},$$

we now define a number I to be the Henstock–Kurzweil integral of f if for every $\varepsilon > 0$ there exists a gauge δ such that whenever P is δ-fine, we have

$$\left| \sum_{P} f - I \right| < \varepsilon.$$

If such an I exists, we say that f is Henstock–Kurzweil integrable on $[a, b]$.

Cousin's theorem states that for every gauge δ, such a δ-fine partition P does exist, so this condition cannot be satisfied vacuously. The Riemann integral can be regarded as the special case where we only allow constant gauges.

Properties

Let f: $[a, b] \to$ R be any function.

If $a < c < b$, then f is Henstock–Kurzweil integrable on $[a, b]$ if and only if it is Henstock–Kurzweil integrable on both $[a, c]$ and $[c, b]$, and then

$$\int_a^b f(x)dx = \int_a^c f(x)dx + \int_c^b f(x)dx.$$

The Henstock–Kurzweil integral is linear, i.e., if f and g are integrable, and α, β are reals, then $\alpha f + \beta g$ is integrable and

$$\int_a^b \alpha f(x) + \beta g(x)dx = \alpha \int_a^b f(x)dx + \beta \int_a^b g(x)dx.$$

If f is Riemann or Lebesgue integrable, then it is also Henstock–Kurzweil integrable, and the values of the integrals are the same. The important Hake's theorem states that

$$\int_a^b f(x)dx = \lim_{c \to b-} \int_a^c f(x)dx$$

whenever either side of the equation exists, and symmetrically for the lower integration bound. This means that if f is "improperly Henstock–Kurzweil integrable", then it is properly Henstock–Kurzweil integrable; in particular, improper Riemann or Lebesgue integrals such as

$$\int_0^1 \frac{\sin(1/x)}{x} dx$$

are also Henstock–Kurzweil integrals. This shows that there is no sense in studying an "improper Henstock–Kurzweil integral" with finite bounds. However, it makes sense to consider improper Henstock–Kurzweil integrals with infinite bounds such as

$$\int_a^{+\infty} f(x)dx := \lim_{b\to+\infty} \int_a^b f(x)dx.$$

For many types of functions the Henstock–Kurzweil integral is no more general than Lebesgue integral. For example, if f is bounded, the following are equivalent:

- f is Henstock–Kurzweil integrable,
- f is Lebesgue integrable,
- f is Lebesgue measurable.

In general, every Henstock–Kurzweil integrable function is measurable, and f is Lebesgue integrable if and only if both f and $|f|$ are Henstock–Kurzweil integrable.

This means that the Henstock–Kurzweil integral can be thought of as a "non-absolutely convergent version of Lebesgue integral".

It also implies that the Henstock–Kurzweil integral satisfies appropriate versions of the monotone convergence theorem (without requiring the functions to be nonnegative) and dominated convergence theorem (where the condition of dominance is loosened to $g(x) \le f_n(x) \le h(x)$ for some integrable g, h).

If F is differentiable everywhere (or with countable many exceptions), the derivative $F2†$ is Henstock–Kurzweil integrable, and its indefinite Henstock–Kurzweil integral is F. (Note that $F2†$ need not be Lebesgue integrable.)

In other words, we obtain a simpler and more satisfactory version of the second fundamental theorem of calculus: each differentiable function is, up to a constant, the integral of its derivative:

$$F(x) - F(a) = \int_a^x F'(t)dt.$$

Conversely, the Lebesgue differentiation theorem continues to holds for the Henstock–Kurzweil integral: if f is Henstock–Kurzweil integrable on $[a, b]$, and

$$F(x) = \int_a^x f(t)dt,$$

then $F'(x) = f(x)$ almost everywhere in $[a, b]$ (in particular, F is almost everywhere differentiable).

The space of all Henstock–Kurzweil-integrable functions is often endowed with the Alexiewicz norm, with respect to which it is barrelled but incomplete.

McShane Integral

Interestingly, Lebesgue integral on a line can also be presented in a similar fashion.

First of all, change of

$$\forall i \;\; u_i - u_{i-1} < \delta(t_i)$$

to

$$\forall i \;\; [u_{i-1}, u_i] \subset U_{\delta(t_i)}(t_i)$$

(here $U_\varepsilon(a)$ is a ε-neighbourhood of a) in the notion of δ-fine partition yields a definition of the Henstock–Kurzweil integral equivalent to the one given above. But after this change we can drop condition

$$t_i \in [u_{i-1}, u_i]$$

and get a definition of *McShane integral*, which is equivalent to the Lebesgue integral.

Chapter 8

Basic Properties of Derivative

We continue our study of classes of functions which are suitably restricted. Again we are passing from the general to the particular. The next most particular class of function we study after the class of continuous functions is the class of differentiable functions. We discuss the definition, show how to get "new functions from old" in what by now is a fairly routine way, and prove that this is a smaller class: that every differentiable function is continuous, but that there are continuous functions that are not differentiable. Informally, the difference is that the graph of a differentiable function may not have any sharp corners in it.

As with continuous functions, our aim is to show that there are many attractive properties which hold for differentiable functions that don't hold in general for continuous functions. One we discuss in some detail is the ease with which certain limits can be evaluated, by introducing l'Hopital's rule. Although we don't prove this, the corresponding results are false for continuous functions. We take the view that much of this material has already been discussed last year, so we move fairly quickly over the basics.

Definition: Let U be an open subset of $\mathbb{R}$, and let f : $U \to \mathbb{R}$. We say that f is differentiable at $a \in U$ iff

$$\lim_{x \to \alpha} \frac{f(x) - f(a)}{x - a}$$

or equivalently,

$$\lim_{h\to 0}\frac{f(a+h)-f(a)}{h}$$

exists.

The limit, if it exists, is written as $f'(a)$.

We say that f is differentiable in U if and only if it is differentiable at each a in U.

Note that the Newton quotient is not defined when $x = a$, nor need it be for the definition to make sense. But the Newton quotient, if it exists, can be extended to be a continuous function at a by defining its value at that point to be $f'(a)$. Note also the emphasis on the existence of the limit. Differentiation is as much about showing the *existence* of the derivative, as *calculating* the value of the derivative.

Example: Let $f(x) = x^3$. Show, directly from the definition, that $f'(a) = 3a^2$.

Solution: This is just another way of asking about particular limits, we must compute

$$\lim_{x\to\alpha}\frac{x^3-a^3}{x-a} = \lim_{x\to\alpha}\frac{(x-a)(x^2+xa+a^2)}{x-a} = \lim_{x\to\alpha}(x^2 + 2xa + a^2) = 3a^2.$$

Exercise: Let $f(x) = \sqrt{x}$. Show, directly from the definition, that $f'(a) = 1/2\sqrt{a}$ when $a \neq 0$. What function do you have to consider in the particular case when $a = 4$?

Just as with continuity, it is impractical to use this definition every time to compute derivatives; we need results showing how to differentiate the usual class of functions, and we assume these are known from last year. Thus we assume the rules for differentiation of sums products and compositions of functions, together with the known derivatives of elementary functions such as sin, cos and tan; their reciprocals sec, *cosec* and cot; and exp and log.

Proposition: Let f and g be differentiable at a, and let k be a constant. Then $k.f$, $f + g$ and fg are differentiable at f. Also, if $g(a) \neq 0$, then f/g is differentiable at a. Let f be differentiable at a, and let g be differentiable at $f(a)$. Then gof is differentiable at a. In addition, the usual rules for calculating these derivatives apply.

Example: Let $f(x) = \tan\left(\frac{x^2 - a^2}{x^2 + a^2}\right)$ for $a \neq 0$. Show that f is differentiable at every point of its domain, and calculate the derivative at each such point.

Solution: This is the same example we considered in. There we showed the domain was the whole of $\mathbb{R}$, and that the function was continuous everywhere. Let $g(x) = \frac{x^2 - a^2}{x^2 + a^2}$. Then g is properly defined for all values of x, and the quotient is differentiable, since each term is, and since $x^2 + a^2 \neq 0$ for any x since $a \neq 0$. Thus f is differentiable using the chain rule since $f = \tan o g$, and we are assuming known that the elementary functions like tan are differentiable.

Finally to actually calculate the derivative, we have:

$$f'(x) = \sec^2\left(\frac{x^2 - a^2}{x^2 + a^2}\right)\frac{\left(x^2 + a^2\right)\cdot 2x - \left(\left(x^2 - a^2\right)\cdot 2x\right)}{\left(x^2 + a^2\right)^2}.$$

$$= \frac{4a^2x}{\left(x^2 + a^2\right)^2}\cdot\sec^2\left(\frac{x^2 - a^2}{x^2 + a^2}\right)$$

Exercise: Let $f(x) = \exp\left(\frac{1 + x^2}{1 - x^2}\right)$. Show that f is differentiable at every point of its domain, and calculate the derivative at each such point. The first point in our study of differentiable functions is that it is *more* restrictive for a function to be differentiable, than simply to be continuous.

Proposition: Let f be differentiable at a. Then f is continuous at a.

Proof: To establish continuity, we must prove that $\lim_{x \to a} f(x) = f(a)$. Since the Newton quotient is known to converge, we have for $x \neq a$,

$$f(x) - f(a) = \frac{f(x) - f(a)}{x - a}.(x - a) \to f'(a).0 \quad \text{as } x \to a.$$

Hence f is continuous at a.

Example: Let $f(x) = |x|$; then f is continuous everywhere, but *not* differentiable at 0.

Solution: We already know from Example that $|x|$ is continuous. We compute the Newton quotient directly; recall t h a t

$|x| = x$ if $x \geq 0$, while $|x| = -x$ if $x < 0$. Thus

$$\lim_{x \to 0+} \frac{f(x) - f(0)}{x - 0} = \lim_{x \to 0+} \frac{x - 0}{x - 0}, \text{ while}$$

$$\lim_{x \to 0-} \frac{f(x) - f(0)}{x - 0} = \lim_{x \to 0-} \frac{-x - 0}{x - 0} = -1.$$

Thus both of the one-sided limits exist, but are unequal, so the limit of the Newton quotient does not exist.

Simple Limits

Our calculus of differentiable functions can be used to compute limits which otherwise prove troublesome.

Proposition: Let f and g be functions such that $f(a) = g(a) = 0$ while $f'(a)$ and $g'(a)$ both exits and $g'(a) \neq 0$. Then

$$\lim_{x \to a} \frac{f(x)}{g(x)} = \frac{f'(a)}{g'(a)}.$$

Proof: Since $f(a) = g(a) = 0$, provided $x \neq a$, we have

$$\frac{f(x)}{g(x)} = \frac{f(x) - f(a)}{g(x) - g(a)} = \frac{f(x) - f(a)}{x - a} \frac{x - a}{g(x) - g(a)}$$

$$\to \frac{f'(a)}{g'(a)} \text{ as } x \to a$$

where the last limit exists, since $g'(a) \neq 0$.

Remark: If $f'(a)$ and $g'(a)$ exist, $\lim_{x\to\alpha} \frac{f(x)}{g(x)}$ computing is easy, since f and g must be continuous at a by Proposition, unless we get an indeterminate form 0/0 or ∞/∞ for the formal quotient.

In fact l'Hopitals rule helps in both cases, although we need to develop stronger forms.

Example: Show that $\lim_{x\to 0} \frac{3x - \sin x}{x} = 2.$

Solution: Note first that we cannot get the result trivially from, since since $g(a) = 0$ and so we get the indeterminate form 0/0. However, we are in a position to apply the simple form of l'Hopital, since $x' = 1 \neq 0$. Applying the rule gives

$$\lim_{x\to 0} \frac{3x - \sin x}{x} = \lim_{x\to 0} \frac{3 - \cos x}{1} = 2.$$

Example: Show that $\lim_{x\to 0} \frac{\sqrt{1+x} - 1}{x} = 1/2.$

Solution: Note again that we cannot get the result trivially from 4.16, since this gives the indeterminate 0/0 form, because $g(a) = 0$. However, we are in a position to apply the simple form of l'Hopital, since $x' = 1 \neq 0$. Applying the rule gives

$$\lim_{x\to 0} \frac{\sqrt{1+x} - 1}{x} - \lim_{x\to 0} \frac{2^{-1}(1+x)^{-1/2}}{1} = 1/2.$$

Example: Show that $\lim_{x\to 0} \frac{\sin x}{x} = 1.$

Solution: This is spurious because we need the limit to calculate the derivative in the first place, but applying l'Hopital certainly gives the result.

Rolle and the Mean Value Theorem

We can combine our definition of derivative with the Intermediate Value Theorem to give a useful result which is in fact the basis of most elementary applications of the differential calculus. Like the results on continuous functions, it is a global result, and so needs continuity and differentiability on a whole interval.

Theorem (Rolle's Theorem): Let f be continuous on $[a, b]$, and differentiable on (a, b), and suppose that $f(a) = f(b)$. Then there is some c with $a < c < b$ such that $f'(c) = 0$.

*Note:*The theorem guarantees that the point c exists somewhere. It gives *no* indication of how to find c. Here is the diagram to make the point geometrically:

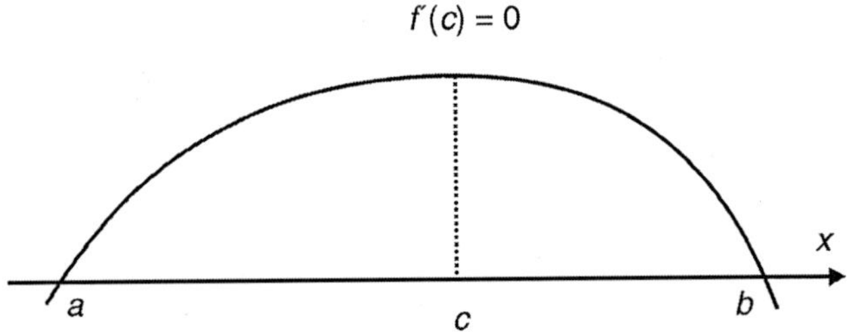

Figure: *If* f *Crosses the Axis twice, somewhere between the two Crossings, the Function is flat. The Accurate Statement of this "Obvious" Observation is Rolle's Theorem.*

***Proof*:** Since f is continuous on the compact interval $[a, b]$, it has both a global maximum and a global minimum. Assume first that the global maximum occurs at an interior point $c \in (a, b)$. In what follows, we pick h small enough so that $c + h$ always lies in (a, b). Then

If $h > 0$, $\dfrac{f(c+h)-f(c)}{h} \leq 0$, and so

$$\lim_{h \to 0+} \frac{f(c+h)-f(c)}{h} \leq 0,$$

since we know the limit exists.

Similarly, if $h < 0$, $\dfrac{f(c+h)-f(c)}{h} \geq 0$,

and so

$$\lim_{h \to 0+} \frac{f(c+h)-f(c)}{h} \geq 0,$$

since we know the limit exists. Combining these, we see that $f'(c) = 0$, and we have the result in this case. A similar argument applies if, instead, the global minimum occurs at the interior point c. The remaining situation occurs if both the global maximum and global minimum occur at end points; since $f(a) = f(b)$, it follows that f is constant, and any $c \in (a, b)$ will do.

Example: Investigate the number of roots of each of the polynomials

$$p(x) = x^3 + 3x + 1 \quad \text{and} \quad q(x) = x^3 - 3x + 1.$$

Solution: Since $p'(x) = 3(x^2 + 1) > 0$ for all $x \in \mathbb{R}$, we see that p has at most one root; for if it had two (or more) roots there would be a root of $p'(x) = 0$ between them by Rolle. Since $p(0) = 1$, while $p(-1) = -3$, there is at least one root by the Intermediate Value Theorem. Hence p has exactly one root. We have $q'(x) = 3(x^2 - 1) = 0$ when $x = \pm 1$. Since $q(-1) = 3$ and $q(1) = -1$, there is a root of q between -1 and 1 by the Intermediate Value Theorem. Looking as $x \to \infty$ and as $x \to -\infty$ shows here are three roots of q.

Taylor's Theorem

We have so far explored the Mean Value theorem, which can be rewritten as

$$f(a + h) = f(a) + hf'(c)$$

where c is some point between a and $a + h$. We have already met the approximation

$$f(a + h) \approx f(a) + hf'(a)$$

when we studied the Newton - Raphson method for solving an equation, and have already observed that the Mean Value Theorem provides a more accurate version of this. Now consider what happens when f is a polynomial of degree n

$$f(x) = a_0 + a_1x + a_2x^2 + ... + a_{n-1}x^{n-1} + a_nx^n.$$

Note that $f(0) = a_0$. Differentiating gives

$$f'(x) = a_1 + 2a_2x + 3a_3x^2 ... + (n-1)a_{n-1}x^{n-2} + na_nx^{n-1},$$

and so $f'(0) = a_1$. Again, we have

$$f''(x) = 2a_2 + 3.2a_3x ... + (n-1)(n-2)a_{n-1}x^{n-3} + n(n-1)a_nx^{n-2},$$

and $f¢¢(0) = 2a_2$. After the next differentiation, we get $f'''(0) = 3!a_3$, while after k differentiations, we get, $f^{(k)}(0) = k!a_k$, provided $k \leq n$. Thus we can rewrite the polynomial, using its value, and the value of its derivatives at 0, as

$$f(x) = f(0) + f'(0)x + \frac{f''(0)}{2!}x^2 + \frac{f'''(0)}{3!}$$

$$x^3 + ... + \frac{f^{(n-1)}(0)}{(n-1)!}x^{n-1} + \frac{f^{(n)}(0)}{n!}x^n.$$

This opens up the possibility of representing more general functions than polynomials in this way, and so getting a generalisation of the Mean Value Theorem.

Taylors Theorem - Lagrange form of Remainder

Theorem: Let f be continuous on $[a, x]$, and assume that each of $f', f'', ..., f^{(n+1)}$ is defined on $[a, x]$. Then we can write

$$f(x) = P_n(x) + R_n(x),$$

where $P_n(x)$, the Taylor polynomial of degree n about a, and $R_n(x)$, the corresponding remainder, are given by

$$P_n(x) = f(a) + f'(a)(x-a) + \frac{f''(a)}{2!}(x-a)^2$$

$$+ \frac{f^{(n)}(a)}{n!}(x-a)^n,$$

$$R_n(x) = \frac{f^{(n+1)}(c)}{(n+1)!}(x-a)^{n+1},$$

where c is some point between a and x. We make no attempt to prove this, although the proof can be done with the tools we have at our disposal. Some quick comments:

1. The theorem is also true for $x < a$; just restate it for the interval $[x, a]$ etc;
2. If $n = 0$, we have $f(x) = f(a) + (x - a)f'(c)$ for some c between a and x; this is a restatement of the Mean Value Theorem;
3. If $n = 1$, we have

$$f(x) = f(a) + (x - a)f'(x) + \frac{f''(c)}{2!}(x - a)^2$$

for some c between a and x; this often called the Second Mean Value Theorem;
4. In general we can restate Taylor's Theorem as

$$f(x) = f(a) + (x - a)f'(x) + \ldots + \frac{f^{(n)}(a)}{2!}(x - a)^n$$
$$+ \frac{f^{(n+1)}(c)}{(n+1)!}(x - a)^{n+1},$$

for some c between a and x;
5. The special case in which $a = 0$ has a special name; it is called Maclaurin's Theorem;
6. Just as with Rolle, or the Mean Value Theorem, there is no useful information about the point c.

We now explore the meaning and content of the theorem with a number of examples.

Example: Find the Taylor polynomial of order n about 0 for $f(x) = e^x$, and write down the corresponding remainder term.

Solution: There is no difficulty here in calculating derivatives clearly $f^{(k)}(x) = e^x$ for all k, and so $f^{(k)}(0) = 1$. Thus by Taylor's theorem,

$$e^x = 1 + x + \frac{x^2}{2!} + \frac{x^2}{2!} + \ldots \frac{x^n}{n!} + \frac{x^{n+1}}{(n+1)!}e^c$$

for some point c between 0 and x. In particular,

$$P_n(x) = 1 + x + \frac{x^2}{2!} + \frac{x^2}{2!} + \ldots \frac{x^n}{n!}$$

and $$R_n(x) = R_n(x) = \frac{x^{n+1}}{(n+1)!} e^c .$$

We can actually say a little more about this example if we recall that x is *fixed.* We have

$$e^x = P_n(x) + R_n(x) = P_n(x) + \frac{x^{n+1}}{(n+1)!} e^c$$

We show that $R_n(x) \to 0$ as $n \to \infty$, so that (again for fixed x), the sequence $P_n(x) \to e^x$ as n. If $x < 0$, $e^c < 1$, while if $x \geq 1$, then since $c < x$, we have $e^c < e^x$. thus

$R_n(x) =$ emax(e, 1)0 as n.

$$\left|R_{n(x)}\right| = \left|\frac{x^{n+1}}{(n+1)!} e^c\right| \leq \frac{|x|^{n+1}}{(n+1)!}$$

max $(e^x, 1) \to 0$ as $n \to \infty$.

We think of the limit of the polynomial as forming a series, the Taylor series for e^x.

We study series (and then Taylor series).

Example: Find the Taylor polynomial of order 1 about a for $f(x) = e^x$, and write down the corresponding remainder term.

Solution: Using the derivatives computed above, by Taylor's theorem,

$$e^x = e^\alpha + (x-a)e^\alpha + \frac{(x-a)^2}{2!} e^c$$

for some point c between a and x. In particular,

$$P_1(x) = e^\alpha + (x-a)e^\alpha \text{ and } R_1(x) = \frac{(x-a)^2}{2!} e^c .$$

Example: Find the Maclaurin polynomial of order $n > 3$ about 0 for $f(x) = (1 + x)^3$, and write down the corresponding remainder term.

Solution: We have

$$f(x) = (1 + x)^3$$
$$f'(x) = 3(1 + x)^2$$
$$f''(x) = 6(1 + x)$$
$$f'''(x) = 6$$
$$f^{(n)}(x) = 0 \text{ if } n > 3.$$

$$f(0) = 1$$

$$f'(x) = 3$$

$$f''(x) = 6$$

$$f'''(x) = 6$$

and so, by Taylor's theorem

$$(1 + x)^3 = 1 + 3x + \frac{6}{2!}x^2 + \frac{6}{3!}x^3,$$

a result we could have got directly, but which is at least reassuring.

Find the Taylor polynomial of order n about 0 for $f(x) = \sin x$, and write down the corresponding remainder term.

Solution: There is no difficulty here in calculating derivatives we have

$$f(x) = \sin x$$
$$f'(x) = \cos x$$
$$f''(x) = -\sin x$$
$$f'''(x) = -\cos x$$
$$f^{(4)}(x) = \sin x \text{ and so on.}$$

$$f(0) = 0$$

$$f'(x) = 1$$

$$f''(x) = 0$$

$$f'''(x) = -1$$

Thus by Taylor's theorem,

$$\sin x = x - \frac{x^3}{3!} + \frac{x^5}{5!} + \ldots + (-1)^{n+1} \frac{x^{2n+1}}{(2n+1)!}$$

Writing down the remainder term isn't particularly useful, but the important point is that

$$| R_{2n+1}(x) | \left| \frac{x^{2n+3}}{(2n+3)!} \right| \to 0 \text{ as } n \to \infty .$$

Exercise: Recall that $\cosh x = \dfrac{e^x + e^{-x}}{2}$, and that sinh,

$$x = \frac{e^x + e^{-x}}{2}.$$

Now check the shape of the following Taylor polynomials:

$$\cos x = 1 - \frac{x^2}{2!} + \frac{x^4}{4!} + \ldots + (-1)^n \frac{x^{2n}}{2n!} + \ldots$$

$$\sinh x = x + \frac{x^3}{3!} + \frac{x^5}{5!} + \ldots + \frac{x^{2n+1}}{(2n+1)!} + \ldots$$

$$\cosh x = 1 + \frac{x^2}{2!} + \frac{x^4}{4!} + \ldots + \frac{x^{2n}}{2n!} + \ldots$$

Example: Find the maximum error in the approximation

$$\sin(x) \sim x - \frac{x^3}{3!} \text{ given that } | x | < 1/2.$$

Solution: We use the Taylor polynomial for sin x of order 4 about 0, together with the corresponding remainder. Thus

$$\sin x = x - \frac{x^3}{3!} + \frac{x^5}{5!}\cos c$$

for some c with $0 < c < 1/2$ or $-1/2 < c < 0$. In any case, since $|x| < 1/2$,

$$\left|\frac{x^5}{5!}\cos c\right| \le \left|\frac{x^5}{5!}\right| \le \left|\frac{1}{2^5.5!}\right| \le \frac{1}{120.32}$$

Warning: The Taylor polynomial always exists, providing f is suitably differentiable. But it need not be useful. Consider the example

$$(x) = \begin{cases} \exp\left(-1/x^2\right) \text{if } x > 0; \\ 0 \qquad \text{if } x \le 0 \end{cases}$$

The interest in f is at 0; it is well behaved everywhere else. It turns out that,

$f(0) = f¢(0) = f¢¢(0) =...= f^{(n)}(0) =...= 0.$

So the Taylor polynomial of degree n for f about 0 is $P_n(x) = 0 + 0x + 0x^2 +...+ 0.x^n = 0$, and so for every n, $R_n(x) = f(x)$. Clearly in this case, P_n tells us nothing useful about the function.

Example: Find the Taylor polynomial of order n about 0 for $f(x) = (1 + x)^a$, and note that this gives a derivation of the binomial theorem. In fact, the remainder $|R_n(x)| \to 0$ as $n \to \infty$, provided $|x| < 1$.

Solution: There is again no difficulty here in calculating derivatives we have

$$f(x) = (1+x)^{\alpha}$$

$$f'(x) = \alpha(1+x)^{\alpha-1}$$

$$f''(x) = \alpha(\alpha-1)(1+x)^{\alpha-2}$$

$$f'''(x) = \alpha(\alpha-1)(\alpha-2)(1+x)^{\alpha-3}$$

and in general

$$f^{(n)}(x)=\alpha(\alpha-1)...(\alpha-n+1)(1+x)^{\alpha-n}$$

$$f(0)=1$$

$$f'(x)=\alpha$$

$$f''(x)=\alpha(\alpha-1)$$

$$f'''(x)=\alpha(\alpha-1)(\alpha-2)$$

$$f^{(n)}=\alpha(\alpha-1)...(\alpha-n+1)$$

$$f(x)=(1+x)^{\alpha}$$

$$f'(x)=\alpha(1+x)^{\alpha-1}$$

$$f''(x)=\alpha(\alpha-1)(1+x)^{\alpha-2}$$

$$f'''(x)=\alpha(\alpha-1)(\alpha-2)(1+x)^{\alpha-3}$$

and in General,

$$f^{(n)}(x)=\alpha(\alpha-1)...(\alpha-n+1)(1+x)^{\alpha-n}$$

$$f(0)=1$$

$$f'(x)=\alpha$$

$$f''(x)=\alpha(\alpha-1)$$

$$f'''(x)=\alpha(\alpha-1)(\alpha-2)$$

$$f^{(n)}=\alpha(\alpha-1)...(\alpha-n+1)$$

Thus by Taylor's theorem,

$$(1+x)^{\alpha}=1+\alpha x-\frac{\alpha(\alpha-1)}{2!}x^2+\frac{\alpha(\alpha-1)(\alpha-2)}{3!}x^3+$$

$$...+\frac{\alpha(\alpha-1)...(\alpha-n+1)}{n!}x^n+...$$

The remainder is not hard to deal with, but we omit the proof; in fact $|R_n(x)\to 0|$ when $n\to\infty$.

Note also that if $\alpha > 0$ is an *integer*, say $\alpha = n$ then $|R_n(x)| = 0$ and $f(x) = P_n(x)$. This is another way to get the Binomial theorem described.

Infinite Series

In this section we return to study a particular kind of sequence, those built by adding up more and more from a given collection of terms. One motivation comes from section, in which we obtained a sequence of approximating polynomials $\{P_n\}$ to a given function. It is more natural to think of adding additional terms to the polynomial, and as such we are studying series. However, there is a closely related sequence; the sequence of partial sums.

Arithmetic and Geometric Series Consider the Sum

$$a + (a + r) + (a + 2r) + \cdots + (a + nr)$$

We are trying to add up the terms in an arithmetic progression. A small amount of notation makes the addition easy.

Let $S_n = a + (a + r) + (a + 2r) + \cdots + (a + nr)$,

$S_n = (a + nr) + (a + (n - 1)r) + \cdots + (a + r) + a$,

so $2S_n = (2a + nr) + (2a + nr) + \cdots + (2a + nr)$, and

$$S_n = (n + 1).\frac{(\alpha + (\alpha + nr))}{2}.$$

Note that if $r > 0$ then $S_n \to \infty$ as $n \to \infty$, while if $r < 0$ then $S_n \to -\infty$ as $n \to \infty$.

We next consider a geometric progression (or series):

Let $S_n = a + ar + ar^2 + \cdots + ar^n$,

$rS_n = ar + ar^2 + ar^3 + \cdots + ar^{n+1}$,

so $(1 - r)\, S_n = a(1 - r^{n+1})$, and

$$S_n = \frac{a(1 - r^{n+1})}{1 - r} \quad \text{if} \quad r \neq 1.$$

Note that if $|r| < 1$, then $S_n \to \dfrac{a}{1-r}$ as $n \to \infty$.

If $r > 1$, say $r = 1 + k$,

$$S_n = \quad a\left(\frac{(1+k)^{n+1}-1}{k}\right),$$

$$> \frac{a}{k}\left(1+(n+1)_{k-1}\right) > a(n+1) \to \infty \text{ if } a > 0.$$

Convergent Series

Definition: Let $\{a_n\}$ be a sequence, and let,

$S_n = a_1 + a_2 + \cdots + a_n$, the nth partial sum.

If $\lim_{n\to\infty} s_n$ exists, we say that $\sum a_n$ is a convergent series, and write $\lim_{n\to\infty} s_n = \sum a_n$.

Thus a series is convergent if and only if it's sequence of partial sums is convergent. The limit of the sequence of partial sums is the sum of the series. A series which is *not* convergent, is a divergent series.

Example: The series $\sum r^n$ is convergent with sum $1/(1 - r)$, provided that $|r| < 1$. For other values of r, the series is divergent; in particular, the series $\sum(-1)^n$ is divergent.

Solution: We noted above that when $|r| < 1$, $S_n \to a/(1 - r)$ as $n \to \infty$; note particular cases;

$$\sum_{n=1}^{\infty} \frac{1}{2^n} = 1 \text{ or equivalently, } \frac{1}{2}+\frac{1}{4}+\frac{1}{8}+\ldots = 1.$$

Example: The sum $\sum \frac{1}{n(n+1)}$ is convergent with sum 1.

Solution: We can compute the partial sums explicitly:

$$S_n = \sum_{k=1}^{n} \frac{1}{k(k+1)} = \sum_{k=1}^{n} \left(\frac{1}{k} - \frac{1}{k+1} \right) = 1 - \frac{1}{n+1} \to 1 \text{ as } n \to \infty$$

Example: The sum $\sum \frac{1}{n}$ is divergent.

Solution: We estimate the partial sums:

$$S_n = \frac{1}{1} + \frac{1}{2} + \left(\frac{1}{3} + \frac{1}{4} \right) + \left(\frac{1}{5} + \ldots + \frac{1}{8} \right) + \left(\frac{1}{9} + \ldots + \frac{1}{15} \right) + \ldots + \frac{1}{n}$$

$$> 1 + \frac{1}{2} + \frac{2}{4} + \frac{4}{8} > 2\frac{1}{2} \text{ if } n \geq 15$$

$$> 1 + \frac{1}{2} + \frac{2}{4} + \frac{4}{8} + \frac{8}{16} > 3$$

if $n \geq 31$

$$\to \infty \text{ as } n \to \infty .$$

Example: The sum is $\sum \frac{1}{n^2}$ convergent. [Actually the sum is $\frac{\pi^2}{6}$, but this is much harder.]

Solution: We estimate the partial sums. Since $1/n^2 > 0$, clearly $\{S_n\}$ is an increasing sequence. We show it is bounded above, whence by the Monotone Convergence Theorem, it is convergent. From the diagram,

$$\frac{1}{2^2} + \frac{1}{3^2} + \ldots + \frac{1}{n^2} = \int_1^n \frac{dx}{x^2}, \text{ and so}$$

$$S_n < 1 + \left[-\frac{1}{x} \right]_1^n \leq 2 - \frac{1}{n}$$

Thus $S_n < 2$ for all n, the sequence of partial sums is bounded above, and the series is convergent.

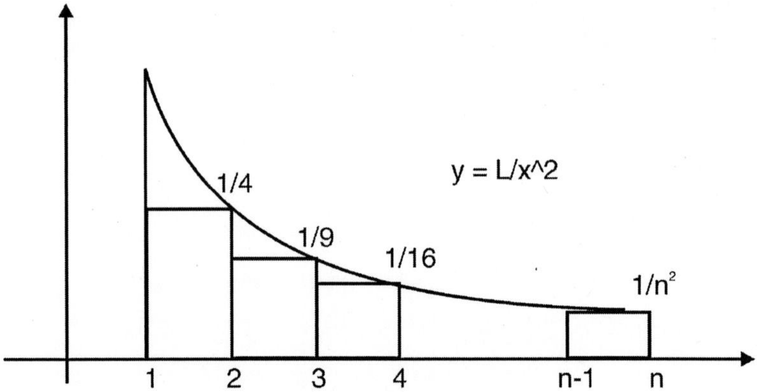

Figure: *Comparing the area under the Curve y = $1/x^2$ with the area of the Rectangles Below the Curve*

Proposition: Let $\sum a_n$ be convergent. Then $a_n \to 0$ as $n \to \infty$.

Proof: Write $l = \lim_{n\to\infty} s_n$, and recall from our work on limits of sequences that $S_{n-1} l \to l$ as $n \to \infty$.

Then,

$$a_n = (a_1 + a_2 + \ldots a_n) - (a_1 + a_2 + \ldots a_{n-1})$$
$$= S_n - S_{n-1} l - l \quad \text{as} \quad n.$$

Example: The sum $\sum \frac{1}{n}$ is divergent (Graphical method).

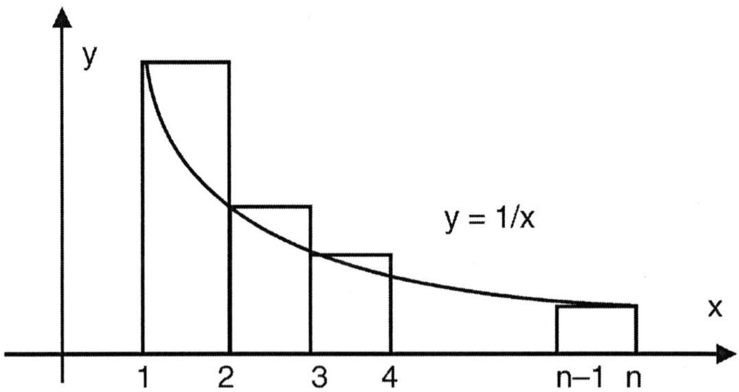

Figure: *Comparing the area under the Curve y = 1/x with the area of the Rectangles above the Curve*

Solution: We estimate the partial sums. Since $1/n > 0$, clearly $\{S_n\}$ is an increasing sequence. We show it is not bounded above, whence by the note after 3.9, the sequence of partial sums $\to \infty$ as $n \to \infty$.

From the diagram,

$$1+\frac{1}{2}+\ldots+\frac{1}{n}>\int_1^n \frac{dx}{x}>\frac{1}{2}+\ldots+\frac{1}{n}$$

Writing $S_n = 1+\frac{1}{2}+\cdots+\frac{1}{n}$, we have $S_n > \log n > S_n - 1$, or equivalently $1 + \log n > S_n > \log n$ for all n. Thus $S_n \to \infty$ and the series is divergent.

Proposition: Let $\sum a_n$ and $\sum b_n$ be convergent. Then $\sum (a_n + b_n)$ and $\sum c.a_n$ are convergent.

Proof: This can be checked easily directly from the definition; it is in effect the same proof that the sum of two convergent sequences is convergent etc

Chapter 9

Discontinuous Function

The simplest discontinuity occurs at those points at which a function has a jump, that is, at which the function has a definite limit as x tends to the point from the right and a definite limit as x tends to the point from the left, while these two limits are different. It does not matter whether or how the function is defined at the point of discontinuity itself.

For example, the function $f(x)$ defined by the equations

$$f(x)=0 \text{ for } x^2>1, f(x)=1 \text{ for}$$

$$x^2>1, f(x)=\frac{1}{2} \text{ for } x^3=1$$

has discontinuities at the points $\xi=1$ and $\xi=-1$. The limits on approaching these points from the right and from the left hand side differ by 1, and the values of the function at these points agree with neither limit, but are equal to the arithmetic mean of the two limits. It may be noted in passing that our function can be represented, using the idea of a limit, by the expression

$$f(x) = \lim_{x\to\infty} \frac{1}{1+x^{2n}}$$

For, if $x < 1$, that is, if x lies in the interval $-1 < < 1$, the numbers x^{2n} will have the limit 0, and the function will have the value 1. However, if $x > 1$, as n increases, x^{2n} will increase beyond all bounds; our function will then have the value 0.

Finally, obviously, for $x^2 = 1$, that is, for $x = +1$ and $x = -1$, the value of the function is.

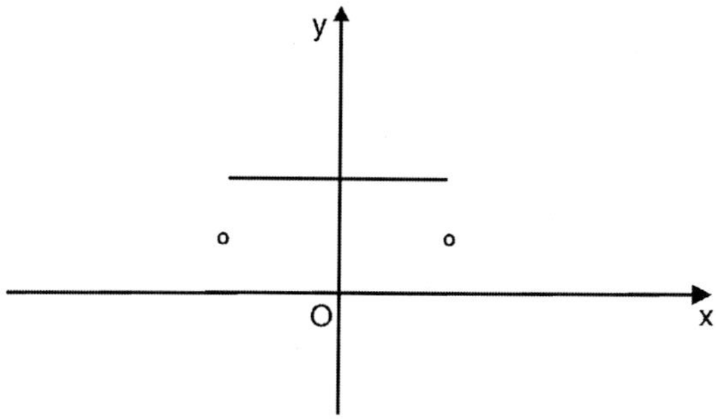

Other curves with jumps are shown in Figures; they represent functions with obvious discontinuities.

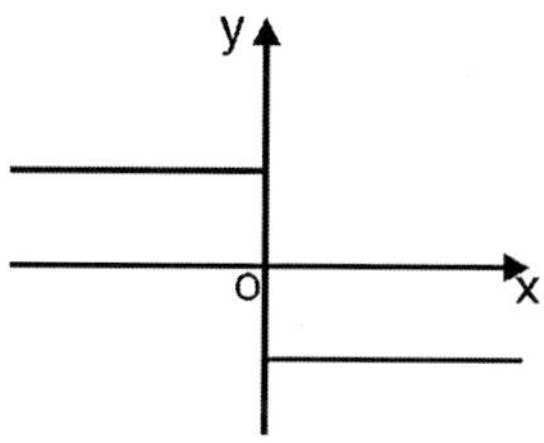

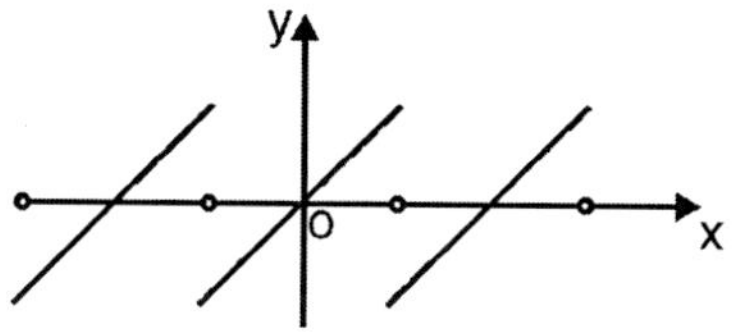

In the case of discontinuities of this kind, both the limits from the right and from the left exist.

We now proceed to the consideration of discontinuities in which this is not the case.

These most important of such discontinuities are the infinite discontinuities or infinities. These are discontinuities which are exhibited by the functions $1/x$ or l/x at the point $\xi = 0$; as x ξ, the absolute value | $f(x)$ | of the function increases beyond all bounds.

In the case of $1/x$, the function increases numerically beyond all bounds through positive and through negative values, respectively, as x approaches the origin from the right and from the left. On the other hand, the function l/x has for $x = 0$ an infinite discontinuity at which its value becomes positive from both sides. The function $y = 1/(x - 1)$, shown in Figure, has infinite discontinuitics both at $x - 1$ and at $x -$ 1. Finally, we shall illustrate by an example another type of discontinuity in which there does not exist a limit from the right or from the left. Consider the function

$$y = \sin\frac{1}{x}$$

defined for all non-zero values of x. This function takes all values between –1 and +1 as the number $1/x$ ranges through the values from $(2n - 1/2)\pi$ to $(2n+1/2)\pi$, no matter what value has n. At the points $x=2/(4n-1)\pi$, the function will have the value –1, at the points $x = 2/(4n+1)\pi$ the value +1.

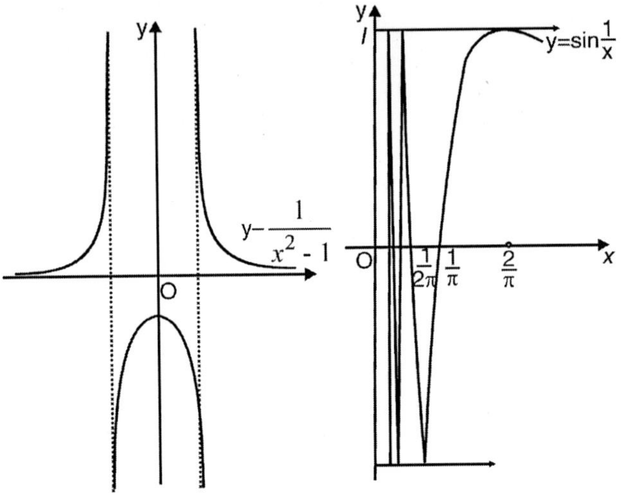

Figure: *Function with Infinite Discontinuities, Oscillating Function with Discontinuity.*

We see from this that the function swings backwards and forwards more rapidly between the values +1 and -1 as x approaches closer and closer to the point 0 and that there occur in the immediate neighbourhood of the point $x = 0$ an infinite number of oscillations. It is interesting to observe that, in contrast to the above example, the function $y = x\sin x/x$ remains continuous at the point $x = 0$, if we assign to it the value 0 at that point. This continuity is due to the fact that, as the origin is approached, the factor x damps the oscillations of the sine.

Yet, in the neighbourhood of the origin, the function $y = x \sin 1/x$ does not change a finite number of times from monotonic increasing to monotonic decreasing.

On the contrary, it oscillates backwards and forwards an infinite number of times, the magnitude of these oscillations becoming as small as we please as the origin is approached. This example shows us that even the simple idea of continuity admits all sorts of remarkable possibilities foreign to our naive intuition. There is one important fact which must be taken into account, if we are to give our ideas greater precision.

It may happen that at a certain point a function is not defined by the original law, as for example at the point $x = 0$ in the last two examples discussed. We have then the right to extend the definition of the function by assigning to it any desired value at such a point. However, in the last example, we can extend the definition in such a way that the function also remains continuous at that point, namely, by setting $y = 0$ when $x = 0$.

This can be done whenever both the limits from the left and from the right exist and are equal to each other; we need then only make the value of the function at the point in question equal to these limits, in order to make the function continuous there. In the case of the function $y=\sin 1/a$, this cannot be done.

Continuous Functions Theorem

In conclusion, we quote the following important, general theorems, the proofs of which follow immediately from the remarks on operations with limits:

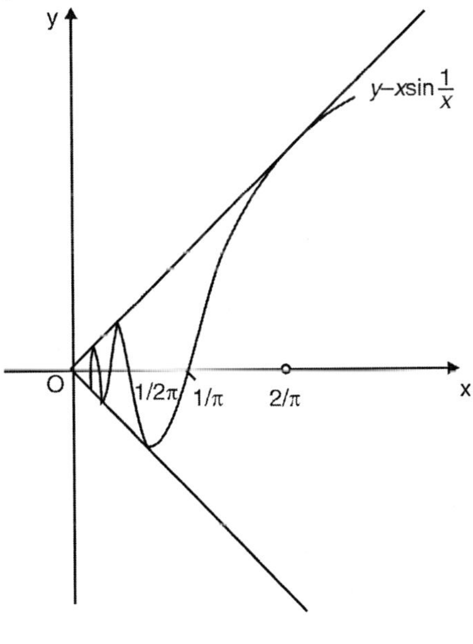

The sum, difference, and product of two continuous functions are themselves continuous. The quotient of two continuous functions is continuous at every point at which the denominator does not vanish.

In particular, it follows that all polynomials and all rational functions are continuous except at the points where the denominator vanishes. The fact that the other elementary functions, such as the trigonometric functions, are continuous will follow naturally from later considerations.

One Sided Limits

Although sometimes we get results directly, it is usually helpful to have a larger range of techniques. We give one here and more in section.

Definition: Say that $\lim\limits_{x \to a-} f(x) = l$, or that f has a limit from the left iff given $\in > 0$, there is some $\delta > 0$ such that whenever $a - \delta < x < a$, then $| f(x) - f(a) | < \in$. There is a similar definition of "limit from the right", writen as $\lim\limits_{x \to a+} f(x) = l$

Example: Define $f(x)$ as follows:

$$f(x) = \begin{cases} 3 - x & \text{if} \, x < 2 \\ 2 & \text{if} \, x < 2; \\ x/2 & \text{if} \, x > 2 \end{cases}$$

Calculate the left and right hand limits of $f(x)$ at 2.

Solution: As $x \to 2-$, $f(x) = 3 - x \to 1+$, so the left hand limit is 1. As $x \to 2+$, $f(x) = x/2 \to 1+$, so the right hand limit is 1. Thus the left and right hand limits agree (and disagree with $f(2)$, so f is not continuous at 2). Note our convention: If $f(x) \to 1$ and always $f(x) \geq 1$ as $x \to 2-$, we say that $f(x)$ tends to 1 from above, and write $f(x) \to 1+$ etc.

Proposition: If $\lim\limits_{x \to a} f(x)$ exists, then both one sided limts exist and are equal. Conversely, if both one sided limits exits and are equal, then $\lim\limits_{x \to a} f(x)$ exists. This splits the problem of finding whether a limit exists into two parts; we can look on either side, and check first that we have a limit, and second, that we get the same answer. For example both 1-sided limits exist, but are not equal. There is now an obvious way of checking continuity.

Continuity Test

Proposition: The function f is continuous at a iff both one sided limits exits and are equal to $f(a)$.

Example: Let $f(x) = \begin{cases} x^2 \text{ for } x \leq 1, \\ x \text{ for } x \geq 1. \end{cases}$

Show that f is continuous at 1.

Solution: We use the above criterion. Note that $f(1) = 1$. Also

$$\lim_{x \to 1-} f(x) = \lim_{x \to 1-} x^2 = 1$$

while
$$\lim_{x \to 1+} f(x) = \lim_{x \to 1+} x = 1 = f(1).$$

so f is continuous at 1.

Example: Let $f(x) = |x|$. Then f is continuous in $\mathbb{R}$.

Solution: Note that if $x < 0$ then $|x| = -x$ and so is continuous, while if $x > 0$, then $|x| = x$ and so also is continuous. It remains to examine the function at 0. From these identifications, we see that,

$$\lim_{x \to 0-} |x| = 0+,$$

while $\lim_{x \to 0+} |x| = 0+$. Since $0+ = 0- = 0 = |0|$, by the $|x|$ is continuous at 0.

Results giving Continuity

Just as for sequences, building continuity directly by calculating limits soon becomes hard work. Instead we give a result that enables us to build new continuous functions from old ones, just as we did for sequences.

Note that if f and g are functions and k is a constant, then $k.f$, $f+g$, fg and (often) f/g are also functions.

Proposition: Let f and g be continuous at a, and let k be a constant. Then $k.f$, $f+g$ and fg are continuous at f. Also, if $g(a) \neq 0$, then f/g is continuous at a.

Proof: We show that $f+g$ is continuous at a. Since, by definition, we have $(f+g)(a) = f(a) + g(a)$, it is enough to show that

$$\lim_{x\to\alpha} \left(f(x)+f(x)\right)\ 4 = f(a) + g(a).$$

Pick $\in > 0$; then there is some δ_1 such that if $|x-a| < \delta_1$, then $|f(x)-f(a)| </2$. Similarly there is some δ_2 such that if $|x-a| < \delta_2$, then $|g(x)-g(a)| < \in/2$. Let $= \min(\delta_1, \delta_2)$, and pick x with $|x-a| < \delta$. Then $|f(x)+g(x)-(f(a)+g(a))| \leq |f(x)-f(a)| + |g(x)-g(a)| < \in/2 + \in/2 = \in$.

This gives the result. The other results are similar, but rather harder; for proofs.

Note: Just as when dealing with sequences, we need to know that f/g is defined in some neighbourhood of a. This can be shown using a very similar proof to the corresponding result for sequences.

Proposition: Let f be continuous at a, and let g be continuous at $f(a)$. Then gof is continuous at a

Proof: Pick $\in > 0$. We must find $\delta > 0$ such that if $|x-a| < \delta$, then $g(f(x)) - g(f(a))| < \in$.

We find δ using the given properties of f and g. Since g is continuous at $f(a)$, there is some $\delta_1 > 0$ such that if $|y-f(a)| < \delta_1$, then $|g(y) - g(f(a))| < \in$.

Now use the fact that f is continuous at a, so there is some $\delta > 0$ such that if $|x-a| < \delta$, then $|f(x)-f(a)| < \delta_1$. Combining these results gives the required inequality.

Infinite Limits

There are many more definitions and results about limits. First one that is close to the sequence definition:

Definition: Say that $\lim_{x \to \infty} f(x) = l$ iff given $\in > 0$, there is some K such that whenever $x > K$, then $| f(x) - l | < \in$.

Example: Evaluate $\lim_{x \to \infty} \frac{x^2 + 3}{3x^2 + 2x + 1}$.

Solution: The idea here should be quite familiar from our sequence work. We use the fact that $1/x \to 0$ as $x \to \infty$. Thus

$$\frac{x^2 + 3}{3x^2 + 2x + 1} = \frac{1 + 3/x^2}{3 + 2/x + 1/x^2} \to \frac{1}{3} \text{ as } x \to \infty .$$

Definition Say that $\lim_{x \to \infty} f(x) = \infty$ iff given $L > 0$, there is some K such that whenever $x > K$, then $f(x) > L$.

The reason for working on proofs from the definition is to be able to check what results of this type are trivially true without having to find it in a book. For example

Proposition: Let $g(x) = 1/f(x)$. Then $g(x) \to 0+$ as x iff $f(x)$ as x. Let $y = 1/x$. Then $y \to 0+$ as $x \to \infty$; conversely, $y \to \infty$ as $x \to 0+$

Proof: Pick $\in > 0$. We show there is some K such that if $x > K$, then $0 < y < \in$; indeed, simply take $K = 1/\in$. The converse is equally trivial.

Closed Interval Continuity

So far our results have followed because f is continuous at a particular point.

In fact we get the best results from assuming rather more. Indeed the results in this section are precisely why we are interested in discussing continuity in the first place.

Although some of the results are "obvious", they only follow from the continuity property, and indeed we present counterexamples whenever that fails. So in order to be able to use the following helpful results, we must first be able to check our functions satisfy the hypothesis of continuity. That task is of course what we have just been concentrating on.

Definition: Say that f is continuous on $[a, b]$ iff f is continuous on (a, b), and if, in addition, $\in \lim_{x \to a+} f(x) = f(a)$, and $\in \lim_{x \to b-} f(x) = f(b)$. We sometimes refer to f as being continuous on a compact interval. Such an f has some very nice properties.

Theorems on Boundedness

Theorem Let f be continuous on the compact interval $[a, b]$. Then there are constants M and m such that $m < f(x) < M$ for all $x \in [a, b]$. In other words, we are guaranteed that the graph of f is bounded both above and below.

Proof: This again uses the completeness of $\mathbb{R}$, and again no proof is offered. Note also that the hypotheses are all needed.

Theorem: Let f be continuous on the compact interval $[a, b]$. Then there are points x_0 and x_1 such that $f(x_0) < f(x) < f(x_1)$ for all $x \in [a, b]$. In other words, we are guaranteed that the graph of f is bounded both above and below, and that these global extrema are actually attained.

Proof: This uses the completeness of $\mathbb{R}$, and follows in part from the previous result.

Consider the function $g(x) = (M - f(x))^{-1}$. This is clearly continuous. If there is some point at which $f(y) = M$, there is nothing to prove; otherwise $g(x)$ is defined everywhere, and is continuous, and hence bounded. This contradicts the fact that M was a least upper boundfor f. We have seen that any polynomial function $P(x)$ satisfies:

$$\lim_{x \to a} P(x) = P(a)$$

for all real numbers a. This property is known as continuity.

Definition: Let $f(x)$ be a function defined on an interval around a. We say that $f(x)$ is continuous at a iff

$$\lim_{x \to a} f(x) = f(a)$$

Otherwise, we say that $f(x)$ is discontinuous at a.

Note that the continuity of $f(x)$ at a means two things:

1. $\lim_{x \to a} f(x)$ exists,
2. And this limit is $f(a)$.

So to be discontinuous at a, means:

1. $\lim_{x \to a} f(x)$ does not exist,
2. or if $\lim_{x \to a} f(x)$ exists, then this limit is not equal to $f(a)$.

Basic Properties of Limits

Theorem. If $f(x)$ and $g(x)$ are continuous at a. Then

1. $f(x) + g(x)$ is continuous at a;
2. $\alpha f(x)$ is continuous at a, where is an arbitrary number;
3. $f(x)g(x)$ is continuous at a;
4. $\frac{f(x)}{g(x)}$ is continuous at a, provided $g(a) \neq 0$;
5. If $f(x)$ is positive, i.e. $f(x) \geq 0$, then $\sqrt{f(x)}$ is continuous at a;
6. If $f(x)$ is continuous at a and $g(x)$ is continuous at $f(a)$, then their composition $g o f(x)$ is continuous at a.

Remark: Many functions are not defined on open intervals. In this case, we can talk about one-sided continuity. Indeed, $f(x)$ is said to be continuous from the left at a iff

$$\lim_{x \to a-} f(x) = f(a)$$

and $f(x)$ is said to be continuous from the right at a iff

$$\lim_{x \to a+} f(x) = f(a)$$

Example: The function $f(x)=\sqrt{x}$ is defined for $x \geq 0$. So we can not talk about left-continuity of *f*(*x*) at 0. But since we conclude that *f*(*x*) is right-continuous at 0. This concept is also important for step-functions.

Example: Consider the function

$$f(x) = \begin{cases} x^3+2 & if \quad x<2 \\ 5 & if \quad x=2 \\ x^2+6 & if \quad x>2 \end{cases}$$

The details are left to the reader to see

$$\lim_{x\to 2-} f(x) = \lim_{x\to 2-} x^3+2=10,$$

and

$$\lim_{x\to 2+} f(x) = \lim_{x\to 2+} x^2+6=10.$$

So we have

$$\lim_{x\to 2} f(x) = 10.$$

Since *f*(2) = 5, then *f*(*x*) is not continuous at 2.

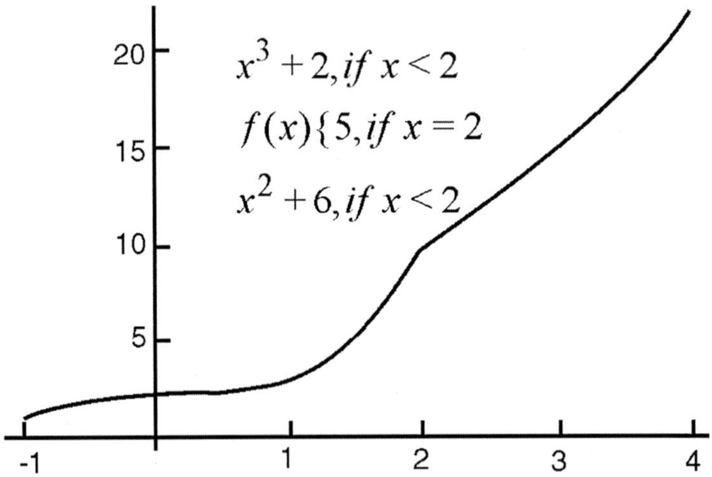

The Intermediate Value Theorem

We can draw the graph of a continuous function without lifting your pencil off the paper. This is made precise by the following result: Intermediate Value Theorem. Let $f(x)$ be a continuous function on the interval $[a, b]$. If $d \in [f(a), f(b)]$, then there is a $c \in [a, b]$ such that $f(c) = d$.

In the case where $f(a) > f(b)$, $[f(a), f(b)]$ is meant to be the same as $[f(b), f(a)]$. Another way to state the Intermediate Value Theorem is to say that the image of a closed interval under a continuous function is a closed interval. We will present an outline of the proof of the Intermediate Value Theorem.

Here is a classical consequence of the Intermediate Value Theorem:

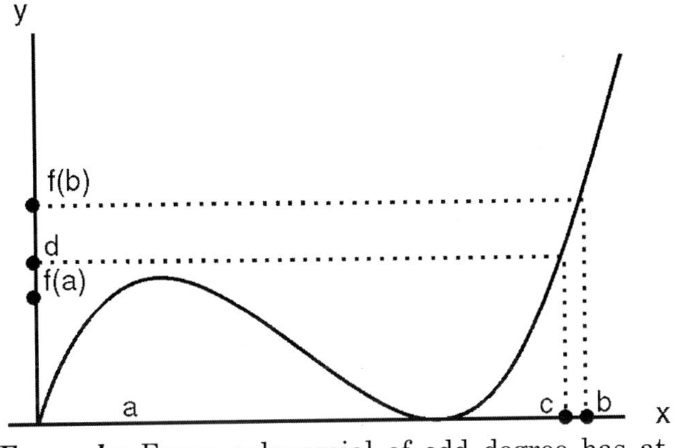

Example: Every polynomial of odd degree has at least one real root. We want to show that if $P(x) = a_n x^n + a_{n-1} x^{n-1} + \cdots + a_1 x + a_0$ is a polynomial with n odd and $a_n \neq 0$, then there is a real number c, such that $P(c) = 0$.

First let me remind you that it follows from the results in previous pages that every polynomial is continuous on the real line. There you also learned that

$$\lim_{x \to \infty} \frac{P(x)}{a_n x^n} = 1 \text{ and } \lim_{x \to -\infty} \frac{P(x)}{a_n x^n} = 1$$

Consequently for $|x|$ large enough, $P(x)$ and $a_n x^n$ have the same sign. But $a_n x^n$ has opposite signs for positive x and negative x. Thus it follows that if $a_n > 0$, there are real numbers $x_0 < x_1$ such that $P(x_0) < 0$ and $P(x_1) > 0$.

Similarly if $a_n < 0$, we can find $x_0 < x_1$ such that $P(x_0) > 0$ and $P(x_1) < 0$. In either case, it now follows directly from the Intermediate Value Theorem that (for $d = 0$) there is a real number $c \in [x_0, x_1]$ with $P(c) = 0$.

The natural question arises whether every function which satisfies the conclusion of the Intermediate Value Theorem must be continuous. Unfortunately, the answer is no and counterexamples are quite messy. The easiest counterexample is the function

$$f(x) = \begin{cases} \sin\left(\frac{1}{x}\right) & ,\text{if } x \neq 0 \\ 0 & ,\text{if } x \neq 0 \end{cases}$$

This function fails to be continuous at $x = 0$. On the other hand, it is not too hard to see that $f(x)$ has the "Intermediate Value Property" even on closed intervals containing $x = 0$.

Intermediate Value Theorem

Theorem: Let f be continuous on the compact interval $[a, b]$, and assume that $f(a) < 0$ and $f(b) > 0$. Then there is some point c with $a < c < b$ such hat $f(c) = 0$.

Proof: We make no attempt to prove this formally, but sketch a proof with a pair of sequences and a repeated bisection argument. It is also noted that each hypothesis is necessary.

Example: Show there is at least one root of the equation $x - \cos x = 0$ in the interval [0, 1].

Proof: Apply the Intermediate Value Theorem to f on the closed interval [0, 1]. The function is continuous on that interval, and $f(0) = -1$, while $f(1) = 1 - \cos(1) > 0$. Thus there is some point $c \in (0, 1)$ such that $f(c) = 0$ as required.

Exercise: Show there is at least one root of the equation $x - e^{-x} = 0$ in the interval (0, 1).

Corollary: Let f be continuous on the compact interval $[a, b]$, and assume there is some constant h such that $f(a) < h$ and $f(b) > h$. Then there is a point c with $a < c < b$ such that $f(c) = h$.

Proof: Apply the Intermediate Value Theorem to f - h on the closed interval [0, 1]. Note that by considering $-f + h$ we can cope with the case when $f(a) > h$ and $f(b) < h$.

Note: This theorem is the reason why continuity is often loosely described as a function you can draw without taking your pen from the paper. As we have seen with $y = 1/x$, this can give an inaccurate idea; it is in fact more akin to connectedness.

The Bisection Method

The Bisection Method at the same time gives a proof of the Intermediate Value Theorem and provides a practical method to find roots of equations. If your calculator can solve equations numerically, it most likely uses a combination of the Bisection Method and the Newton-Raphson Method.

Recall the statement of the Intermediate Value Theorem: Let $f(x)$ be a continuous function on the interval $[a, b]$. If $d \in [f(a), f(b)]$, then there is a $c \in [a, b]$ such that $f(c) = d$.

By replacing $f(x)$ by $f(x)$ - d, we may assume that $d = 0$; it then suffices to obtain the following version: Let $f(x)$ be a continuous function on the interval $\in [a, b]$.

If $f(a)$ and $f(b)$ have opposite signs, then there is a $c \in [a, b]$ such that $f(c) = 0$. Here is an outline of its proof: Let's assume that $f(a) < 0$, while $f(b) > 0$, the other case being handled similarly. Set $a_0 = a$ and $b_0 = b$.

Now consider the midpoint,

$$m_0 = \frac{a_0 + b_0}{2},$$

and evaluate $f(m_0)$. If $f(m_0) < 0$, set $a_1 = m_0$ and $b_1 = b_0$. If $f(m_0) > 0$, set $a_1 = a_0$ and $b_1 = m_0$. (If $f(m_0)=0$, you're already done.) What situation are we in now? $f(a_1)$ and $f(b_1)$ still have opposite signs, but the length of the interval $[a_1, b_1]$ is only half of the length of the original interval $[a_0, b_0]$. Note also that $a_0 \leq a_1$ and that $b_0 \geq b_1$.

The second step: Consider the midpoint,

$$m_1 = \frac{a_1 + b_1}{2},$$

and evaluate $f(m_1)$. If $f(m_1) < 0$, set $a_2 = m_1$ and $b_2 = b_1$. If $f(m_1) > 0$, set $a_2 = a_1$ and $b_2 = m_1$. (If $f(m_1)=0$, you're already done.) What situation are we in now? $f(a_2)$ and $f(b_2)$ still have opposite signs, but the length of the interval $[a_2, b_2]$ is only a quarter of the length of the original interval $[a_0, b_0]$. Note also that $a_0 \leq a_1 \leq a_2$ and that

$$4b_0 \geq b_1 \geq b_2.$$

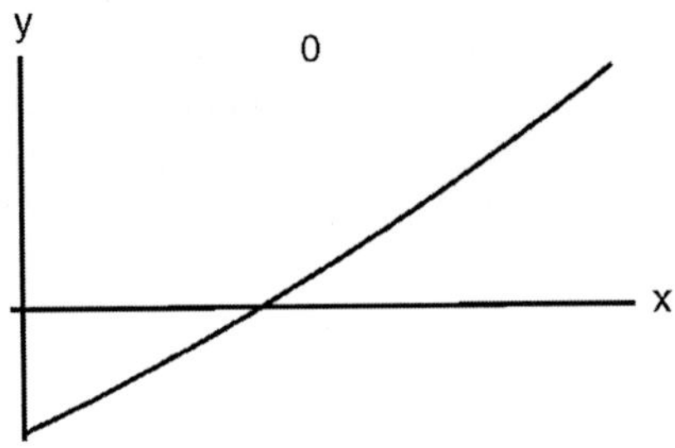

Figure: *The Red Line Shows the Interval $[a_n, b_n]$.*

Continuing in this fashion we construct by induction two sequences,

$$(a_n)_{n=1}^{\infty} \text{ and } (b_n)_{n=1}^{\infty}$$

with the following properties:

1. (a_n) is an increasing sequence, (b_n) is a decreasing sequence.
2. $a_n \leq b_n$ for all n.
3. $f(a_n) < 0$ for all n, $f(b_n) > 0$ for all n.
4. $b_n - a_n = 2^{-n}(b - a)$ for all n.

It follows from the first two properties that the sequences (a_n) and (b_n) converge; set

$$\lim_{x \to \infty} a_n = a, \quad \lim_{x \to \infty} b_n = b.$$

The third property and the continuity of the function $f(x)$ imply that $f(a) \leq 0$ and that $f(b) \geq 0$. The crucial observation is the fact that the fourth property implies that $a = b$. Consequently, $f(a) = f(b) = 0$, and we are done.

Example: Let's compute numerical approximations for the $\sqrt{2}$ with the help of the bisection method. We set $f(x) = x^2$

– 2. Let us start with an interval of length one: $a_0 = 1$ and $b_1 = 2$. Note that $f(a_0) = f(1) = -1 < 0$, and $f(b_0) = f(2) = 3 > 0$. Here are the first 20 applications of the bisection algorithm:

n	a_n	a_n
0	1.	2.
1	1.	1.5
2	1.25	1.5
3	1.375	1.5
4	1.375	1.4375
5	1.40625	1.4375
6	1.40625	1.4375
7	1.4140625	1.421875
8	1.4140625	1.41796875
9	1.4140625	1.416015625
10	1.4140625	1.415039063
11	1.4140625	1.414550781
12	1.4140625	1.414306641
13	1.4148457	1.414306641
14	1.4148457	1.414245605
15	1.4148457	1.414215088
16	1.41499829	1.414215088
17	1.414207458	1.414215088
18	1.414211273	1.414215088
19	1.414213181	1.414215088
20	1.414213181	1.414214134

A comparison of the Bisection Method and the Newton-Raphson Method. The Newton-Raphson Method is often much faster than the Bisection Method. In the last example, we started with an interval of length 1. After 10 steps, the interval $[a_{10}, b_{10}]$ has length 1/1024. Consequently every 10 steps of the Bisection Method will give us about 3 digits more accuracy - that is rather slow. The Newton-Raphson Method can be unreliable: If the algorithm encounters a point x where $f'(x) = 0$, it crashes; if it encounters points where the derivative is very close to 0, it will become very unreliable.

The Bisection Method on the other hand will always work, once you have found starting points a and b where the function takes opposite signs.

The Derivative

The concept of Derivative is at the core of Calculus and modern mathematics. The definition of the derivative can be approached in two different ways. One is geometrical and the other one is physical.

Historically there was a fight between mathematicians which of the two illustrates the concept of the derivative best and which one is more useful.

We will not dwell on this and will introduce both concepts. Our emphasis will be on the use of the derivative as a tool.

Geometrical Concept of Derivative

Consider a function $y = f(x)$ and its graph. Recall that the graph of a function is a set of points (that is $(x,f(x))$ for x's from the domain of the function f).

We may draw the graph in a plane with a horizontal axis (usually called the x-axis) and a vertical axis (usually called the y-axis).

Fix a point on the graph, say $(x_0, f(x_0))$. If the graph as a geometric figure is "nice" (i.e. smooth) around this point, it is natural to ask whether one can find the equation of the straight line "touching" the graph at that point.

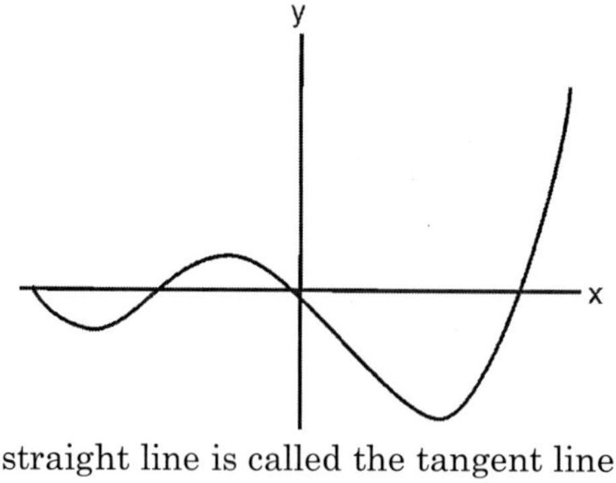

Such a straight line is called the tangent line at the point in question. The concept of tangent may be viewed in a more general framework.

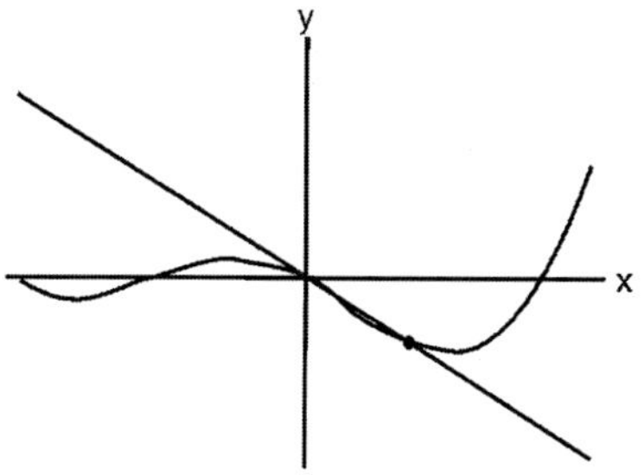

One way to find the tangent line is to consider points $(x, f(x))$ on the graph, where x is very close to x_0. Then draw the straight-line joining both points:

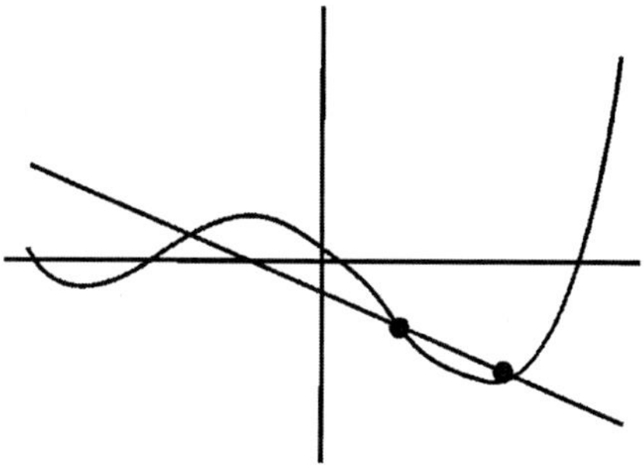

As you can see, when x get closer and closer to x_0, the lines get closer and closer to the tangent line. Since all these lines pass through the point $(x_0, f(x_0))$, their equations will be determined by finding their slope: The slope of the line passing through the points $(x_0, f(x_0))$ and $(x, f(x))$ (where $x \neq x_0$) is given by,

$$m(x) = \frac{f(x) - f(x_0)}{x - x_0}$$

The tangent itself will have a slope m, which is very close to $m(x)$ when x itself is very close to x_0. This is the concept of limit once again! In other words, we have

$$m = \lim_{x \to x_0} m(x) = \lim_{x \to x_0} \frac{f(x) - f(x_0)}{x - x_0}$$

So the equation of the tangent line is,

$$y - f(x_0) = m(x - x_0).$$

Notation. Writing "m" for the slope of the tangent line does not carry enough information; we want to keep track of the function $f(x)$ and the point x_0 in our notation. The common notation used is

$$m = f'(x_0).$$

In this case, the equation of the tangent line becomes

$$y - f(x_0) = f'(x_0)\,(x - x_0)$$

where

$$f'(x_0) = \lim_{x \to x_0} \frac{f(x) - f(x_0)}{x - x_0}$$

One last remark: Sometimes it is more convenient to compute limits when the variable approaches 0. One way to do that is to make a translation along the x-axis. Indeed, if we set $h = x - x_0$, we get

$$f'(x_0) = \lim_{x \to x_0} \frac{f(x) - f(x_0)}{x - x_0}$$

$$\lim_{h \to 0} \frac{f(x_0 + h) - f(x_0)}{h}$$

Chapter 10

Derivative Equations

Rate of Change

The derivative of a function at a chosen input value describes the best linear approximation of the function near that input value. For a real-valued function of a single real variable, the derivative at a point equals the slope of the tangent line to the graph of the function at that point. In higher dimensions, the derivative of a function at a point is a linear transformation called the linearization. The process of finding a derivative is called differentiation.

The fundamental theorem of calculus states that differentiation is the reverse process to integration. The mathematical relationship between an independent and dependent dimension. The relationship expressed in the form of a mathematical function defines a situation for a given set of conditions and properties. Since we allowed the independent dimension to change, it was no longer a fixed quantity but became a variable. The graph of the function reflected a horizontal change of the independent variable with a vertical change in the dependent dimension. This graph allows us to visualize how a situation changes with respect to a change in the conditions. To find out how much the function, $f(x)$, changes between two points x_1 and x_2, we simply enter in the two values for the independent variable x and then calculate the difference between the dependent variable, f, for those given conditions.

Remember that a variable is nothing more than a dimension that is allowed to change or take on any value. Thus, from x_1 to x_2, the change in the independent variable, referred to as Δx is:

$$\Delta x = x_1 - x_2$$

The corresponding change in the dependent variable , referred to as Δf is:

$\Delta f = f(x_2) - f(x_1)\Delta x$ refers to an interval over which we are analyzing the change in the dependent dimension, f.. The second point x_2 can be written in terms of the first point x_1 plus the change in the variable, Δx

$$x_2 - x_1 = \Delta x$$
$$\therefore x_2 = x_1 + \Delta x$$

Therefore the change in the dependent variable over an interval x_1x_2 from $\Delta f = f(x_2) - f(x_1)$ which can also be written as:

$$\Delta f = f(x_1 + \Delta x) - f(x_1)$$

where $\Delta x = x_2 - x_1$

For example the change in the function $f(x) = 3 \cdot x^3\, x = 3$ to $x = 5$, where

Δx is $5 - 3 = 2$ is:

$$\Delta f = f(x_1 + \Delta x) - f(x_1)$$
$$\Delta f = f(3 + 2) - f(3)$$
$$\Delta f = 3.5^3 - 3.3^3 = 375 - 81$$
$$\Delta f = 294$$

Average Rate of Change

We have learned that a change in the independent variable is defined as $\Delta x = x_2 - x_1$, and the corresponding change in the dependent variable over this interval is $\Delta f = f(x_1 + \Delta x) - f(x_1)$ The question we now must ask ourselves is how can we measure the relative change of the dependent variable with respect to the independent variable? In other words how can we calculate how much more or less Δf changed compared to Δx.

To calculate how much more $f(x)$ changed over an interval from x_1, x_2, we simply divide the change in f over the change in x for the interval. Thus we divide, Δf by the interval over we which are evaluating it, Δx which is equal to $x_2 - x_1$. Thus the relative change of f with respect to x over an interval Δx is defined as:

$$\frac{\Delta f}{\Delta x} = \frac{f(x_2) - f(x_1)}{x_2 - x_1}$$

While this expression may seem rather simple, it does require some explanation. By dividing the change in f by the change in x what we are doing is calculating how much more f changed for a given change in x. For example in the function, $f(x) = 3.x^3$, when x changed from 3 to 5, f changed from 81 to 375. Over this interval of from $x = 3$ to $x = 5$, the Δf was 294.

Thus the relative change in f with respect to a change in the independent variable x is:

$$\frac{\Delta f}{\Delta x} = \frac{f(x_2) - f(x_1)}{x_2 - x_1}$$

The value of 147 tells us that f changes 147 times faster than x over that interval of Δx from $x = 3$ to $x = 5$ only. Thus for each unit change in x, $\Delta x = 1$, the corresponding change in f is 147. We can therefore define the rate of change of a function with respect to its independent variable to be:

$$\frac{\Delta f}{\Delta x} = \frac{f(x_1 + \Delta x) - f(x_1)}{\Delta x}$$

The value, called the rate of change of the function, refers to how much more or less $f(x)$ changes for a unit change in x. Itis only valid is for the interval under consideration, $\Delta x = x_2 - x_1$.

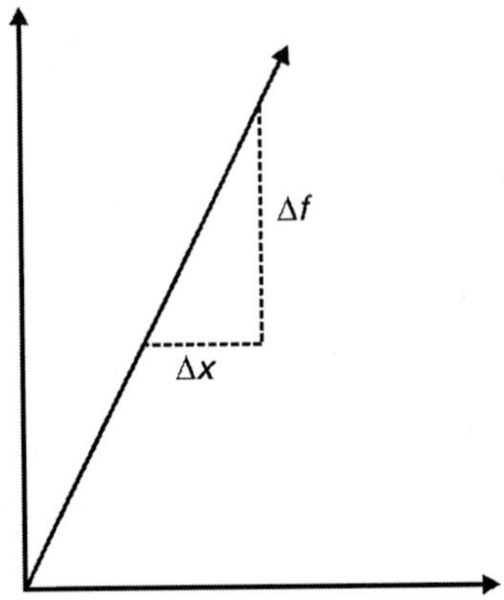

Another way of understanding what rate of change of a function means is to look at the steepness of the line connecting the two endpoints of the interval under consideration.

As Δf increases, the steepness of the line connecting the two endpoints will increase.

Thus, the greater the rate of change of the function, the greater its slope or steepness over the interval under consideration.

Since slope and rate of change are synonymous, then how is rate of change defined for functions whose graphs do not have constant slopes?

For example, from $x = 9$ to $x = 12$ of $f(x) = 3.x^3$, the change in f is 2997. Thus the rate of change of the function over the interval is:

$$\frac{\Delta f}{\Delta x} = \frac{2997}{3} = 999$$

This value is significantly higher than the rate of change calculated for the previous interval from $x = 3$ to $x = 5$.

We can only conclude that the rate of change or slope of the graph must be increasing and is not constant over an interval $\Delta x = x_2 - x_1$.

Instantaneous Rate of Change

In this section we will take a much closer look at rate of change and see how we can define the instantaneous rate of change of a function at any point or value of the independent variable.

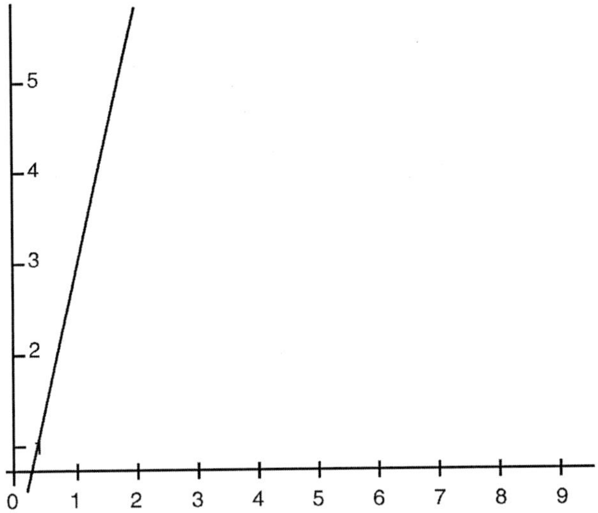

Let us begin with a study of the simple $f(x) = 3x$.

Functions of the form $f(x) = cx$, where c is a constant, express direct relationships. This is because the value of the function, *f*, is a constant multiple or fraction of the independent variable.

Thus f is said to be directly proportional to x. The graph of the function $f(x) = 3x$ looks like. From $x = 1$ to $x = 3$, $\Delta x = 2$, and $\Delta f = 9 - 3 = 6$. Thus, over this interval of $\Delta x = 2$, Δf equals 6. The average rate at which f changed with respect to x is by definition, $\frac{\Delta f}{\Delta x} = \frac{6}{2} = 3$. For each unit change in x, the change in f is 3. This tells us that f is changing three times faster that x is changing over the interval from $x = 1$ to $x = 3$. We can now look at the interval from $x = 2$ to $x = 4$ where Δx equals 2.

$$\Delta f = f(x_1 + \Delta x) - f(x_1)$$
$$\Delta f = f(2+2) - f(2)$$
$$\Delta f = 3.(4) - 3.(2)$$
$$\Delta f = 6$$

Thus

$$\frac{\Delta f}{\Delta x} = \frac{f(x_1 + \Delta x) - f(x_1)}{\Delta x} = \frac{6}{2} = 3$$

Over this interval, the rate of change is the same constant, 3. This leads us to conclude that the rate of change of the function over any interval is a constant, 3. This can be proven by the definition of rate of change:

$$f(x) = 3x$$
$$f(x + \Delta x) = 3(x + \Delta x)$$
$$\frac{\Delta f}{\Delta x} = \frac{f(x + \Delta x) - f(x)}{\Delta x} = \frac{3\cancel{x} + 3\Delta x - 3\cancel{x}}{\Delta x}$$
$$\frac{\Delta f}{\Delta x} = \frac{3\cancel{\Delta x}}{\cancel{\Delta x}} = 3$$

The rate of change of the direct function *f*(*x*) = *cx* is c and is constant over an interval, D*x*. The graph of a function *f*(*x*) = *cx* is therefore an increasing straight line with a constant slope or steepness equal to c. For this reason functions of the form *f*(*x*) = *cx* are also called linear functions since their graphs are straight lines with a constant rate of change.

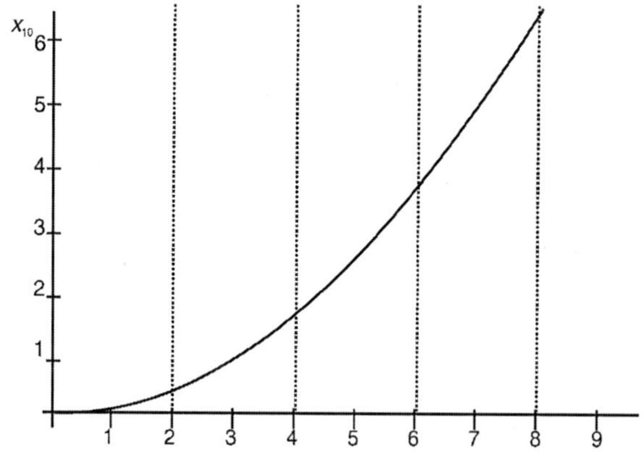

On the other hand, the graph of the function $f(x) = x^2$ is not a straight line. Unlike a line, the rate of change of *f*(*x*) is not the same constant over any interval D*x*. To define the rate of change for the function we will have to derive a more precise way of defining rate of change of function. To do this we will analyze the function over small intervals of Dx

The first point to notice is that the rate of change of the function varies with x.

When x is small, f does not change that much as compared to how much it changes when x is large. We can conclude that the rate of change of f with respect to x, is not constant over any interval, Dx, but varies with x.

To begin our analysis let us divide up the graph into intervals of Dx = *2* and study what is happening in each of these intervals separately. From $x = 0$ to $x = 2$, *Dx* = 2, the change in f, D*f, is f*(2) – *f*(0)= 4. From $x = 6$ to $x = 8$, Dx = 2, the change in *f,* D*f, is f*(8) – *f*(6) = 28.

Clearly, as x increases the rate of change of the function is increasing and is not a constant as in our study of the line where the function changed at the same rate as x increased.

Returning back to the graph of $f(x) = x^2$; we calculated that from $x = 0$ *to* $x = 2$ the change in f was 4. We can then conclude that from $x = 0$ to $x = 2$ the average rate of change of the function over that particular interval is 2 or:

$$\frac{\Delta f}{\Delta x} = \frac{4}{2} = 2$$

Remember it is called average because this rate of change is only valid from $x = 0$ to $x = 2$. Now let us consider the interval from $x = 2$ to $x = 4$ where once again the change in x, Dx, is 2. The change in f of the graph is equal to:

$$\begin{aligned}\Delta f &= f(x + \Delta x) - f(x) \\ &= f(2+2) - f(2) \\ &= f(4) - f(2) \\ &= 16 - 4 \\ &= 12\end{aligned}$$

Thus the average rate of change of the function over this interval is equal to

$$\frac{\Delta f}{\Delta x} = \frac{12}{2} = 6$$

This value is greater than the value we observed from $x = 0$ to $x = 2$. This implies that from $x = 2$ to $x = 4$, f(x) is increasing at a greater rate than from $x = 0$ to $x = 2$. This is despite the fact that in both cases the D$x = 2$. The rate of change is therefore not constant over the but is increasing with x. When we say the rate of change of $f(x)$ from $x = 2$ to $x = 4$ is 6, it is only the average value for the given interval since it assumes rate of change is constant over that interval. Let us move on to the next interval of $x = 4$ to $x = 6$. Once again the change in x or Dx equals 2 but the corresponding change in $f(x)$ is not the same as before.

$$\begin{aligned}\Delta f &= f(x+\Delta x) - f(x)\\ &= (x+\Delta x)^2 - (x)^2\\ &= x^2 + 2x\Delta x + \Delta x^2 - x^2\\ &= 2x\Delta x + \Delta x^2\end{aligned}$$

For x equal to 4 and Dx = 2, the change in the function over this interval is

$$\Delta f = 2x\Delta x + \Delta x^2$$

$$\Delta f = 2(4)(2) + (2)^2$$

$$\Delta f = 20$$

Note that this corresponds to the same value we would get by:

$$\begin{aligned}\Delta f &= f(6) - f(4)\\ &= 6^2 - 4^2\\ &= 36 - 16\\ &= 20\end{aligned}$$

The average rate of change over this interval is therefore:

$$\frac{\Delta f}{\Delta x} = \frac{20}{2} = 10$$

This is still larger than the rate of change for the previous interval which was 6. Remember rate of change , by definition, refers to how much the function changes with respect to a change in the independent variable. The steepness or slope of the line over the interval provides a geometric understanding for this concept.

The average rate of changes calculated over each interval can be used to approximate the graph of $f(x)$ from $x = 0$ *to* $x = 8$.

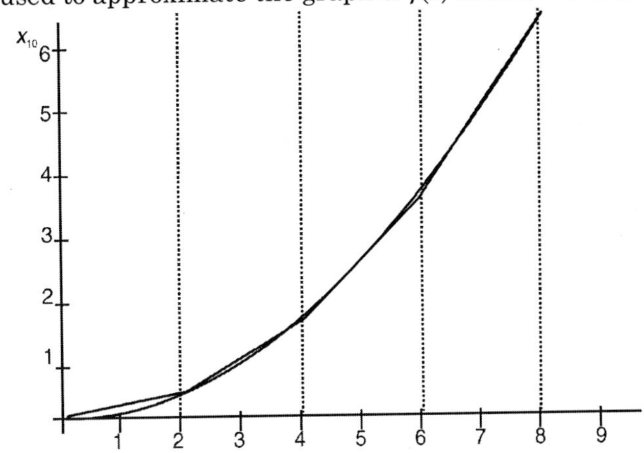

This roughly corresponds to the original graph, but as we can see the rate of change or slope is not constant through out the interval from $x = 0$ *to* $x = 8$, but increases as x-increases. In order to get more accurate answers we need to reduce our interval of Dx = 2 to a much smaller one.

The idea being we need to analyze our graph over a small interval, Dx, to see what exactly is going on at each instant the function is changing. Here is where we begin our study of Calculus. We break down and freeze a changing situation into an infinite series of actions and analyzing what is going on in each individual actions.

Basic Properties of Derivative

We continue our study of classes of functions which are suitably restricted. Again we are passing from the general to the particular. The next most particular class of function we study after the class of continuous functions is the class of differentiable functions. We discuss the definition, show how to get "new functions from old" in what by now is a fairly routine way, and prove that this is a smaller class: that every differentiable function is continuous, but that there are continuous functions that are not differentiable. Informally, the difference is that the graph of a differentiable function may not have any sharp corners in it.

As with continuous functions, our aim is to show that there are many attractive properties which hold for differentiable functions that don't hold in general for continuous functions.

One we discuss in some detail is the ease with which certain limits can be evaluated, by introducing l'Hopital's rule. Although we don't prove this, the corresponding results are false for continuous functions.

We take the view that much of this material has already been discussed last year, so we move fairly quickly over the basics.

Definition: Let U be an open subset of $\mathbb{R}$, and let f : $U \to \mathbb{R}$. We say that f is differentiable at $a \in U$ iff

$$\lim_{x \to \alpha} \frac{f(x) - f(a)}{x - a}$$

or equivalently,

$$\lim_{h \to 0} \frac{f(a+h) - f(a)}{h}$$

exists.

The limit, if it exists, is written as $f'(a)$.

We say that f is differentiable in U if and only if it is differentiable at each a in U.

Note that the Newton quotient is not defined when $x = a$, nor need it be for the definition to make sense. But the Newton quotient, if it exists, can be extended to be a continuous function at a by defining its value at that point to be $f'(a)$. Note also the emphasis on the existence of the limit.

Differentiation is as much about showing the *existence* of the derivative, as *calculating* the value of the derivative.

Example: Let $f(x) = x^3$. Show, directly from the definition, that $f'(a) = 3a^2$.

Solution: This is just another way of asking about particular limits, we must compute

$$\lim_{x \to \alpha} \frac{x^3 - a^3}{x - a} = \lim_{x \to \alpha} \frac{(x-a)(x^2 + xa + a^2)}{x - a} = \lim_{x \to \alpha} (x^2 + 2xa + a^2) = 3a^2.$$

Exercise: Let $f(x) = \sqrt{x}$. Show, directly from the definition, that $f'(a) = 1/2\sqrt{a}$ when $a \neq 0$. What function do you have to consider in the particular case when $a = 4$?

Just as with continuity, it is impractical to use this definition every time to compute derivatives; we need results showing how to differentiate the usual class of functions, and we assume these are known from last year. Thus we assume the rules for differentiation of sums products and compositions of functions, together with the known derivatives of elementary functions such as sin, cos and tan; their reciprocals sec, *cosec* and cot; and exp and log.

Proposition: Let f and g be differentiable at a, and let k be a constant. Then $k.f$, $f + g$ and fg are differentiable at f. Also, if $g(a) \neq 0$, then f/g is differentiable at a. Let f be differentiable at a, and let g be differentiable at $f(a)$. Then gof is differentiable at a. In addition, the usual rules for calculating these derivatives apply.

Example: Let $f(x) = \tan\left(\frac{x^2 - a^2}{x^2 + a^2}\right)$ for $a \neq 0$. Show that f is differentiable at every point of its domain, and calculate the derivative at each such point.

Solution: This is the same example we considered in. There we showed the domain was the whole of $\mathbb{R}$, and that the function was continuous everywhere. Let $g(x) = \frac{x^2 - a^2}{x^2 + a^2}$. Then g is properly defined for all values of x, and the quotient is differentiable, since each term is, and since $x^2 + a^2 \neq 0$ for any x since $a \neq 0$. Thus f is differentiable using the chain rule since $f = \tan o g$, and we are assuming known that the elementary functions like tan are differentiable.

Finally to actually calculate the derivative, we have:

$$f'(x) = \sec^2\left(\frac{x^2-a^2}{x^2+a^2}\right)\frac{\left(x^2+a^2\right)\cdot 2x-\left(\left(x^2-a^2\right)\cdot 2x\right)}{\left(x^2+a^2\right)^2}.$$

$$= \frac{4a^2x}{\left(x^2+a^2\right)^2}\cdot\sec^2\left(\frac{x^2-a^2}{x^2+a^2}\right)$$

Exercise: Let $f(x) = \exp\left(\frac{1+x^2}{1-x^2}\right)$. Show that f is differentiable at every point of its domain, and calculate the derivative at each such point. The first point in our study of differentiable functions is that it is *more* restrictive for a function to be differentiable, than simply to be continuous.

Proposition: Let f be differentiable at a. Then f is continuous at a.

Proof: To establish continuity, we must prove that $\lim\limits_{x \to a} f(x) = f(a)$. Since the Newton quotient is known to converge, we have for $x \neq a$,

$$f(x) - f(a) = \frac{f(x)-f(a)}{x-a}.(x-a) \to f'(a).0 \quad \text{as } x \to a.$$

Hence f is continuous at a.

Example: Let $f(x) = |x|$; then f is continuous everywhere, but *not* differentiable at 0.

Solution: We already know from Example that $|x|$ is continuous. We compute the Newton quotient directly; recall that $|x| = x$ if $x \geq 0$, while $|x| = -x$ if $x < 0$. Thus

$$\lim_{x \to 0+}\frac{f(x)-f(0)}{x-0} = \lim_{x \to 0+}\frac{x-0}{x-0}, \text{ while}$$

$$\lim_{x \to 0-} \frac{f(x) - f(0)}{x - 0} = \lim_{x \to 0-} \frac{-x - 0}{x - 0} = -1.$$

Thus both of the one-sided limits exist, but are unequal, so the limit of the Newton quotient does not exist.

Simple Limits

Our calculus of differentiable functions can be used to compute limits which otherwise prove troublesome.

Proposition: Let f and g be functions such that $f(a) = g(a) = 0$ while $f'(a)$ and $g'(a)$ both exits and $g'(a) \neq 0$. Then

$$\lim_{x \to a} \frac{f(x)}{g(x)} = \frac{f'(a)}{g'(a)}.$$

Proof: Since $f(a) = g(a) = 0$, provided $x \neq a$, we have

$$\frac{f(x)}{g(x)} = \frac{f(x) - f(a)}{g(x) - g(a)} = \frac{f(x) - f(a)}{x - a} \frac{x - a}{g(x) - g(a)}$$

$$\to \frac{f'(a)}{g'(a)} \text{ as } x \to a$$

where the last limit exists, since $g'(a) \neq 0$.

Remark: If $f'(a)$ and $g'(a)$ exist, $\lim_{x \to \alpha} \frac{f(x)}{g(x)}$ computing is easy, since f and g must be continuous at a by Proposition, unless we get an indeterminate form 0/0 or ∞/∞ for the formal quotient. In fact l'Hopitals rule helps in both cases, although we need to develop stronger forms.

Example: Show that $\lim_{x \to 0} \frac{3x - \sin x}{x} = 2.$

Solution: Note first that we cannot get the result trivially from, since since $g(a) = 0$ and so we get the indeterminate form 0/0. However, we are in a position to apply the simple form of l'Hopital, since $x' = 1 \neq 0$. Applying the rule gives

$$\lim_{x\to 0}\frac{3x-\sin x}{x} = \lim_{x\to 0}\frac{3-\cos x}{1} = 2.$$

Example: Show that $\lim_{x\to 0}\frac{\sqrt{1+x}-1}{x}$ =1/2.

Solution: Note again that we cannot get the result trivially from 4.16, since this gives the indeterminate 0/0 form, because $g(a) = 0$. However, we are in a position to apply the simple form of l'Hopital, since $x' = 1 \neq 0$. Applying the rule gives

$$\lim_{x\to 0}\frac{\sqrt{1+x}-1}{x} = \lim_{x\to 0}\frac{2^{-1}(1+x)^{-1/2}}{1} = 1/2.$$

Example: Show that $\lim_{x\to 0}\frac{\sin x}{x} = 1.$

Solution: This is spurious because we need the limit to calculate the derivative in the first place, but applying l'Hopital certainly gives the result.

Rolle and the Mean Value Theorem

We can combine our definition of derivative with the Intermediate Value Theorem to give a useful result which is in fact the basis of most elementary applications of the differential calculus. Like the results on continuous functions, it is a global result, and so needs continuity and differentiability on a whole interval.

Theorem (Rolle's Theorem): Let f be continuous on $[a, b]$, and differentiable on (a, b), and suppose that $f(a) = f(b)$. Then there is some c with $a < c < b$ such that $f'(c) = 0$.

Proof: Since f is continuous on the compact interval $[a, b]$, it has both a global maximum and a global minimum. Assume first that the global maximum occurs at an interior point $c \in (a, b)$. In what follows, we pick h small enough so that $c + h$ always lies in (a, b). Then

If $h > 0$, $\frac{f(c+h)-f(c)}{h} \leq 0$, and so

$$\lim_{h\to 0+}\frac{f(c+h)-f(c)}{h} \leq 0,$$

since we know the limit exists.

Similarly, if $h < 0$, $\frac{f(c+h)-f(c)}{h} \geq 0$,

and so

$$\lim_{h \to 0+} \frac{f(c+h)-f(c)}{h} \geq 0,$$

since we know the limit exists. Combining these, we see that $f'(c) = 0$, and we have the result in this case. A similar argument applies if, instead, the global minimum occurs at the interior point c. The remaining situation occurs if both the global maximum and global minimum occur at end points; since $f(a) = f(b)$, it follows that f is constant, and any $c \in (a, b)$ will do.

Example: Investigate the number of roots of each of the polynomials

$$p(x) = x^3 + 3x + 1 \quad \text{and} \quad q(x) = x^3 - 3x + 1.$$

Solution: Since $p'(x) = 3(x^2 + 1) > 0$ for all $x \in \mathbb{R}$, we see that p has at most one root; for if it had two (or more) roots there would be a root of $p'(x) = 0$ between them by Rolle. Since $p(0) = 1$, while $p(-1) = -3$, there is at least one root by the Intermediate Value Theorem.

Hence p has exactly one root. We have $q'(x) = 3(x^2 - 1) = 0$ when $x = \pm 1$. Since $q(-1) = 3$ and $q(1) = -1$, there is a root of q between -1 and 1 by the Intermediate Value Theorem. Looking as $x \to \infty$ and as $x \to -\infty$ shows here are three roots of q.

The Mean Value Theorem

Theorem: Let f be continuous on $[a, b]$, and differentiable on (a, b). Then there is some c with $a < c < b$ such that

$$\frac{f(b)-f(a)}{b-a} = f'(c)$$

or equivalently $f(b) = f(a) + (b - a)f'(c)$.

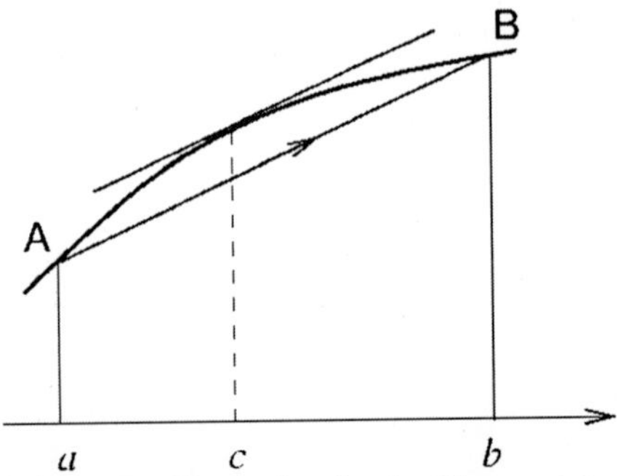

Figure: *Somewhere Inside a chord, the Tangent to f will be Parallel to the chord.*

The Accurate Statement of this Common-sense Observation is the Mean Value Theorem.

Proof: We apply Rolle to a suitable function; let

$$h(x) = f(b) - f(x) - \frac{f(b) - f(a)}{b - a}(b - x).$$

Then h is continuous on the interval $[a, b]$, since f is, and in the same way, it is differentiable on the open interval (a, b). Also, $f(b) = 0$ and $f(a) = 0$. We can thus apply Rolle's theorem to h to deduce there is some point c with $a < c < b$ such that $h'(c) = 0$. Thus we have

$$0 = h'(c) = -f'(c) + \frac{f(b) - f(a)}{b - a},$$

which is the required result.

Example: The function f satisfies $f'(x) = \dfrac{1}{5 - x^2}$ and $f(0) = 2$. Use the Mean Value theorem to estimate f.

Solution: We first estimate the given derivative for values of x satisfying $0 < x < 1$. Since for such x, we have $0 < x^2 < 1$, and so $4 < 5 - x^2 < 5$. Inverting we see that

$$\frac{1}{5} < f'(x) < \frac{1}{4} \text{ when } 0 < x < 1.$$

Now apply the Mean Value theorem to f on the interval $[0, 1]$ to obtain some c with $0 < c < 1$ such that $f(1) - f(0) = f'(c)$. From the given value of $f(0)$, we see that $2.2 < f(1) < 2.25$

Exercise: The function f satisfies $f¢(x)\frac{1}{5+\sin x}=$ and $f(0) = 0$. Use the Mean Value theorem to estimate f (p/2). Note the "common sense" description of what we have done. If the derivative doesn't change much, the function will behave linearly. Note also that this gives meaning to the approximation

$$f(a+h) \approx f(a) + hf'(a).$$

We now see that the accurate version of this replaces $f'(a)$ by $f'(c)$ for some c between a and $a + h$.

The Cauchy Mean Value Theorem

Theorem: Let f and g be both continuous on $[a, b]$ and differentiable on (a, b). Then there is some point c with $a < c < b$ such that

$$g'(c)(f(b) - f(a)) = f'(c)(g(b) - g(a)).$$

In particular, whenever the quotients make sense, we have

$$\frac{f(b)-f(a)}{g(b)-g(a)} = \frac{f'(c)}{g'(c)}$$

Proof. Let $h(x) = f(x)(g(b) - g(a)) - g(x)(f(b) - f(a))$, and apply Rolle's theorem exactly as we did for the Mean Value Theorem. Note first that since both f and g are continuous on $[a, b]$, and differentiable on (a, b), it follows that h has these properties.

Also $h(a) = f(a)g(b) - g(a)f(b) = h(b)$. Thus we may apply Rolle to h, to deduce there is some point c with $a < c < b$ such that $h'(c) = 0$. But

$$h'(c) = f'(c)(g(b) - g(a)) - g'(c)(f(b) - f(a))$$

Thus

$$f'(c)(g(b) - g(a)) = g'(c)(f(b) - f(a))$$

This is one form of the Cauchy Mean Value Theorem for f and g. If $g'(c) \neq 0$ for any possible c, then the Mean Value theorem shows that $g(b) - g(a) \neq 0$, and so we can divide the above result to get

$$\frac{f(b)-f(a)}{g(b)-g(a)} = \frac{f'(c)}{g'(c)},$$

giving a second form of the result.

Note: Taking $g(x) = x$ recovers the Mean Value Theorem.

l'Hopital Revisited

We can get a much more useful form of l'Hopital's rule using the Cauchy Mean Value Theorem, directly from the definition of the derivative.

l'Hopital's Rule: General form

***Proposition*:** Let f and g be functions such that $f(a) = g(a) = 0$, and suppose that f and g are differentiable on an open interval I containing a, and that $g'(x) \neq 0$, except perhaps at a. Then

$$\lim_{x \to \alpha} \frac{f(x)}{g(x)} = \lim_{x \to \alpha} \frac{f'(x)}{g'(x)},$$

provided the second limit exists.

***Proof*:** Pick $x > a$ and apply the Cauchy Mean Value Theorem to the interval $[a, x]$, to find c with $a < c < x$ such that

$$\frac{f(x)}{g(x)} = \frac{f(x)-f(a)}{g(x)-g(a)}$$

$$= \frac{f'(c)}{g'(c)}$$

Then $\lim_{x \to a+} \frac{f(x)}{g(x)} = \lim_{c \to a+} \frac{f'(c)}{g'(c)} = \lim_{x \to a+} \frac{f'(x)}{g'(x)}$, since we know the actual limit (not just the one sided limit) exists. Now repeat with $x < a$ to get the result.

Example: Evaluate $\lim_{x \to 0} \frac{1-\cos x}{x^2} = \frac{1}{2}$.

Solution: We have

$$\lim_{x \to 0} \frac{1-\cos x}{x^2} = \lim_{x \to 0} \frac{\sin x}{2x} = \frac{1}{2}$$

where the use of l'Hopital is justified since the second limit exists. Note that you can't differentiate top and bottom again, and still expect to get the correct answer; one of the hypotheses of l'Hopital is that the first quotient is of the 0/0 form.

Example: Use l'Hopital to establish the following:

$$\lim_{x\to 0} \frac{\sqrt{1+x}-1-x/2}{x^2} = \frac{1}{8}$$

Solution: We have

$$\lim_{x\to 0} \frac{\sqrt{1+x}-1-x/2}{x^2} = \lim_{x\to 0} \frac{(1/2)(1+x)^{-1/2}-1/2}{2x}$$
$$=$$
$$\lim_{x\to 0} \frac{(1/2)(-1/2)(1+x)^{-3/2}}{2} = \frac{-1}{8}$$

The use of l'Hopital is justified the second time, since the third limit exists; since we now know the second limit exists, the use of l'Hopital is justified the first time.

Infinite Limits

We can use Proposition to get results about infinite limits.

Example: Evaluate $\lim_{x\to\infty} x\log\left(1+\frac{1}{x}\right)$

Solution: $\lim_{x\to\infty} x\log\left(1+\frac{1}{x}\right) = \lim_{x\to\infty} \frac{\log(1+y)}{y}$ writing $y =$ $1/x$,

$$= \lim_{y\to 0} \frac{\log(1+y)}{y} = \lim_{y\to 0} \frac{\log(1+y)}{y} - 1$$

The last step is valid, since the final limit exists by l'Hopital; note also that this gives another way of finding $a_n = (1 + 1/n)^n$.

l'Hopital's Rule: Infinite Limits

Proposition: Let f and g be functions such that $\lim_{x\to\infty} f(x) = \lim_{x\to\infty} g(x) = \infty$, and suppose

that $\lim_{x\to\infty} \frac{f'(x)}{g'(x)}$ exists. Then

$$\lim_{x\to\infty}\frac{f(x)}{g(x)}=\lim_{x\to\infty}\frac{f'(x)}{g'(x)}$$

Proof: Pick $\in > 0$ and choose a such that

$$\lim_{x\to\infty}\left|\frac{f'(x)}{g'(x)}\right|-l<\in \quad \text{for all } x>a.$$

Then pick K such that if $x > K$, then $g(x) - g(a) \neq 0$. By Cauchy,

$$\frac{f'(c)}{g'(c)}=\frac{f(x)-f(a)}{g(x)-g(a)} \quad \text{for all } x>K.$$

Note that although c depends on x, we always have $c > a$. Then

$$\frac{f(x)}{g(x)}=\frac{f(x)-f(a)}{g(x)-g(a)}\cdot\frac{f(x)}{f(x)-f(a)}\cdot\frac{g(x)-g(a)}{g(x)}.$$

Chapter 11

Rational Integral

Integrals Involving Roots

In this section we're going to look at an integration technique that can be useful for *some* integrals with roots in them.

Let's look at a couple of examples to see another technique that can be used on occasion to help with these integrals.

Example: Evaluate the following integral. $\int \frac{x+2}{\sqrt[3]{x-3}}\,dx$

Solution: Sometimes when faced with an integral that contains a root we can use the following substitution to simplify the integral into a form that can be easily worked with.

$$u = \sqrt[3]{x-3}$$

So, instead of letting u be the stuff under the radical as we often did in Calculus I we let u be the whole radical.

Now, there will be a little more work here since we will also need to know what x is so we can substitute in for that in the numerator and so we can compute the differential, dx.

This is easy enough to get however. Just solve the substitution for x as follows,

$$x = u^3 + 3 \qquad dx = 3u^2\,du$$

Using this substitution the integral is now,

$$\int \frac{(u^3+3)+2}{u} 3u^2\,du = \int 3u^4 + 15u\,du$$
$$= \frac{3}{5}u^5 + \frac{15}{2}u^2 + c$$
$$= \frac{3}{5}(x-3)^{\frac{5}{3}} + \frac{15}{2}(x-3)^{\frac{2}{3}} + c$$

So, sometimes, when an integral contains the root $\sqrt[n]{g(x)}$ the substitution,

$$u = \sqrt[n]{g(x)}$$

can be used to simplify the integral into a form that we can deal with. Let's take a look at another example real quick.

Example: Evaluate the following integral.

$$\int \frac{2}{x - 3\sqrt{x+10}}\,dx$$

Solution: We'll do the same thing we did in the previous example. Here's the substitution and the extra work we'll need to do to get x in terms of u.

$$u = \sqrt{x+10} \qquad x = u^2 - 10 \qquad dx = 2u\,du$$

With this substitution the integral is,

$$\int \frac{2}{x - 3\sqrt{x+10}}\,dx = \int \frac{2}{u^2 - 10 - 3u}(2u)du = \int \frac{4u}{u^2 - 3u - 10}\,du$$

This integral can now be done with partial fractions.

$$\frac{4u}{(u-5)(u+2)} = \frac{A}{u-5} + \frac{B}{u+2}$$

Setting numerators equal gives,

$$4u = A(u+2) + B(u-5)$$

Picking value of u gives the coefficients.

$$u = -2 \qquad -8 = B(-7) \qquad B = \frac{8}{7}$$

$$u = 5 \qquad 20 = A(7) \qquad A = \frac{20}{7}$$

The integral is then,

$$\int \frac{2}{x-3\sqrt{x+10}}\,dx = \int \frac{\frac{20}{7}}{u-5} + \frac{\frac{8}{7}}{u-2}\,du$$

$$= \frac{20}{7}\ln|u-5| + \frac{8}{7}\ln|u+2| + c$$

$$= \frac{20}{7}\ln|\sqrt{x+10}-5| + \frac{8}{7}\ln|\sqrt{x+10}+2| + c$$

Integrals Involving Quadratics

To this point we've seen quite a few integrals that involve quadratics. A couple of examples are,

$$\int \frac{x}{x^2 \pm a}\,dx = \frac{1}{2}\ln|x^2 \pm a| + c \qquad \int \frac{1}{x^2+a^2}\,dx = \frac{1}{2}\tan^{-1}\left(\frac{x}{a}\right)$$

We also saw that integrals involving, $\sqrt{b^2x^2-a^2}$, $\sqrt{a^2-b^2x^2}$ and $\sqrt{a^2+b^2x^2}$ could be done with a trig substitution. Notice however that all of these integrals were missing an x term. They all consist of a quadratic term and a constant. Some integrals involving general quadratics are easy enough to do. For instance, the following integral can be done with a quick substitution.

$$\int \frac{2x+3}{4x^3+12x}\,dx = \frac{1}{4}\int \frac{1}{u}\,du \quad (u = 4x^2+12x-1 \quad du = 4(2x+3)dx)$$

$$= \frac{1}{4}\ln|4x^2+12x-1| + c$$

Some integrals with quadratics can be done with partial fractions. For instance,

$$\int \frac{10x-6}{3x^2+16x+5}\,dx = \int \frac{4}{x+5} - \frac{2}{3x+1}\,dx = 4\ln|x+5| - \frac{2}{3}\ln|3x+1| + c$$

Unfortunately, these methods won't work on a lot of integrals. A simple substitution will only work if the numerator is a constant multiple of the derivative of the denominator and partial fractions will only work if the denominator can be factored. This section is how to deal with integrals involving quadratics when the techniques that we've looked at to this point simply won't work. Back in the Trig Substitution section we saw how to deal with square roots that had a general quadratic in them. Let's take a quick look at another one like that since the idea involved in doing that kind integral is exactly what we are going to need for the other integrals in this section.

Example: Evaluate the following integral.

$$\int \sqrt{x^2+4x+5}\, dx$$

Solution: Recall from the Trig Substitution section that in order to do a trig substitution here we first needed to complete the square on the quadratic. This gives,

$$x^2+4x+5 = x^2+4x+4-4+5 = (x+2)^2+1$$

After completing the square the integral becomes,

$$\int \sqrt{x^2+4x+5} dx = \int \sqrt{(x+2)^2+1}\, dx$$

Upon doing this we can identify the trig substitution that we need. Here it is,

$$x+2=\tan\theta \qquad x=\tan\theta-2 \qquad dx=\sec^2\theta d\theta$$

$$\sqrt{(x+2)^2+1}=\sqrt{\tan^2\theta+1}=\sqrt{\sec^2\theta}=|\sec\theta|=\sec\theta$$

Recall that since we are doing an indefinite integral we can drop the absolute value bars. Using this substitution the integral becomes,

$$\int \sqrt{x^2+4x+5}\, dx = \int \sec^3\theta d\theta$$

$$= \frac{1}{2}(\sec\theta\tan\theta + \ln|\sec\theta+\tan\theta|)+c$$

We can finish the integral out with the following right triangle.

$$\tan\theta = \frac{x+2}{1} \qquad \sec\theta = \frac{\sqrt{x^2 4x+5}}{1} = \sqrt{x^2+4x+5}$$

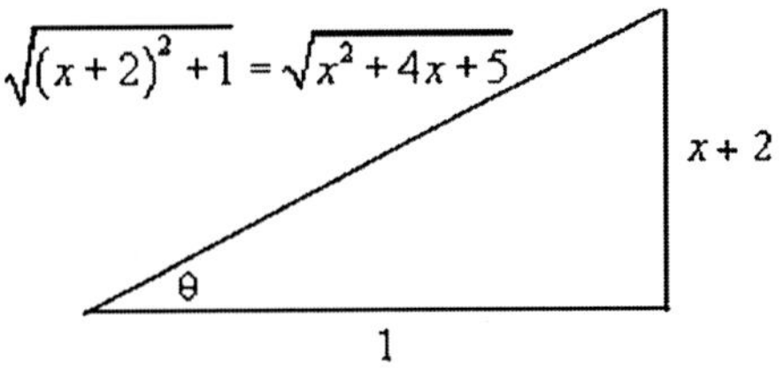

$$\int \sqrt{x^2+4x+5}\,dx = \frac{1}{2}\left((x+2)\sqrt{x^2+4x+5} + \ln\left|x+2+\sqrt{x^2+4x+5}\,\right|\right) + c$$

So, by completing the square we were able to take an integral that had a general quadratic in it and convert it into a form that allowed use a known integration technique.

Let's do a quick review of completing the square before proceeding. Here is the general completing the square formula that we'll use.

$$x^2+bx+c = x^2+bx+\left(\frac{b}{2}\right)^2-\left(\frac{b}{2}\right)^2+c = \left(x+\frac{b}{2}\right)^2+c-\frac{b^2}{4}$$

This will always take a general quadratic and write it in terms of a squared term and a constant term. Recall as well that in order to do this we must have a coefficient of one in front of the x^2.

If not we'll need to factor out the coefficient before completing the square. In other words,

$$ax^2+bx+c = a\left(\underbrace{x^2+\frac{b}{a}x+\frac{c}{a}}_{\substack{\text{complete the}\\ \text{square on that}}}\right)$$

Now, let's see how completing the square can be used to do integrals that we aren't able to do at this point.

Example: Evaluate the following integral.

$$\int \frac{1}{2x^2-3x+2}\,dx$$

Solution: Okay, this doesn't factor so partial fractions just won't work on this.

Likewise, since the numerator is just "1" we can't use the substitution. $u = 2x^2 - 3x + 8$. So, let's see what happens if we complete the square on the denominator.

$$2x^2 - 3x + 2 = 2\left(x^2 - \frac{3}{2}x + 1\right)$$
$$= 2\left(x^2 - \frac{3}{2}x + \frac{9}{16} - \frac{9}{16} + 1\right)$$
$$2\left(\left(x - \frac{3}{4}\right)^2 + \frac{7}{16}\right)$$

With this the integral is,

$$\int \frac{1}{2x^2 - 3x + 2}\,dx = \frac{1}{2}\int \frac{1}{\left(x - \frac{3}{4}\right)^2 + \frac{7}{16}}\,dx$$

Now this may not seem like all that great of a change. However, notice that we can now use the following substitution.

$$u = x - \frac{3}{4} \qquad\qquad du = dx$$

and the integral is now,

$$\int \frac{1}{2x^2 - 3x + 2}\,dx = \frac{1}{2}\int \frac{1}{u^2 + \frac{7}{16}}\,du$$

We can now see that this is an inverse tangent! So, using the formula from above we get,

$$\int \frac{1}{2x^2 - 3x + 2}\,dx = \frac{1}{2}\left(\frac{4}{\sqrt{7}}\right)\tan^{-1}\left(\frac{4u}{\sqrt{7}}\right) + c$$
$$= \frac{2}{\sqrt{7}}\tan^{-1}\left(\frac{4x - 3}{\sqrt{7}}\right) + c$$

Example: Evaluate the following integral.

$$\int \frac{3x - 1}{x^2 + 10x + 28}\,dx$$

Solution: This example is a little different from the previous one. In this case we do have an x in the numerator however the numerator still isn't a multiple of the derivative of the denominator.

So, let's again complete the square on the denominator and see what we get,

$$x^2 + 10x + 28 = x^2 + 10x + 25 - 25 + 28 = (x+5)^2 + 3$$

Upon completing the square the integral becomes,

$$\int \frac{3x-1}{x^2+10x+28}dx = \int \frac{3x-1}{(x+5)^2+3}dx$$

At this point we can use the same type of substitution that we did in the previous example. The only real difference is that we'll need to make sure that we plug the substitution back into the numerator as well.

$$u = x + 5 \quad x = u - 5 \quad dx = du$$

$$\begin{aligned}\int \frac{3x-1}{x^2+10x+28}dx &= \int \frac{3(u-5)-1}{u^2+3}dx \\ &= \int \frac{3u}{u^2+3} - \frac{16}{u^2+3}du \\ &= \frac{3}{2}\ln|u^2+3| - \frac{16}{\sqrt{3}}\tan^{-1}\left(\frac{u}{\sqrt{3}}\right) + c \\ &= \frac{3}{2}\ln|(x+5)^2+3| - \frac{16}{\sqrt{3}}\tan^{-1}\left(\frac{x+5}{\sqrt{3}}\right) + c\end{aligned}$$

So, in general when dealing with an integral in the form,

$$\int \frac{Ax+B}{dx^2+bx+c}dx$$

Here we are going to assume that the denominator doesn't factor and the numerator isn't a constant multiple of the derivative of the denominator. In these cases we complete the square on the denominator and then do a substitution that will yield an inverse tangent and/or a logarithm depending on the exact form of the numerator.

Let's now take a look at a couple of integrals that are in the same general form except the denominator will also be raised to a power. In other words, let's look at integrals in the form,

$$\int \frac{Ax+B}{(ax^2+bx+c)^{"}}dx$$

Example: Evaluate the following integral.

$$\int \frac{x}{(x^2-6x+11)^3}dx$$

Solution: For the most part this integral will work the same as the previous two with one exception that will occur down the road. So, let's start by completing the square on the quadratic in the denominator.

$$x^2-6x+11-x^2--6x+9-9+11=(x-3)^2+2$$

The integral is then,

$$\int \frac{x}{(x^2-6x+11)^3}dx=\int \frac{x}{\left[(x-3)^2+2\right]}dx$$

Now, we will use the same substitution that we've used to this point in the previous two examples.

$$u=x-3 \qquad x=u+3 \qquad dx=du$$

$$\int \frac{x}{(x^2-6x+11)^3}dx=\int \frac{u+3}{(u^2+2)^3}du$$
$$=\int \frac{u}{(u^2+2)^3}du+\int \frac{3}{(u^2+2)}du$$

Now, here is where the differences start cropping up. The first integral can be done with the substitution $v = u^2 + 2$ and isn't too difficult.

The second integral however, can't be done with the substitution used on the first integral and it isn't an inverse tangent.

It turns out that a trig substitution will work nicely on the second integral and it will be the same as we did when we had square roots in the problem.

$$u=\sqrt{2}\tan\theta \qquad du=\sqrt{2}\sec^2\theta d\theta$$

With these two substitutions the integrals become,

$$\int \frac{x}{(x^2-6x+11)^3}dx = \frac{1}{2}\int \frac{1}{v^3}dv + \int \frac{3}{(2\tan^2\theta+2)^2}(\sqrt{2}\sec^2\theta)d\theta$$

$$= \frac{1}{4}\frac{1}{v^2} + \int \frac{3\sqrt{2}\sec^2\theta}{8(\tan^2\theta+1)^3}d\theta$$

$$= -\frac{1}{4}\frac{1}{(u^2+2)^2} + \frac{3\sqrt{2}}{8}\int \frac{\sec^2\theta}{(\sec^2\theta)^3}d\theta$$

$$= \frac{1}{4}\frac{1}{\left((x-3)^2+2\right)} + \frac{3\sqrt{2}}{8}\int \frac{1}{\sec^4\theta}d\theta$$

$$= \frac{1}{4}\frac{1}{\left((x-3)^2+2\right)} + \frac{3\sqrt{2}}{8}\int \cos^4\theta d\theta$$

Okay, at this point we've got two options for the remaining integral.

We can either use the ideas we learned in the section about integrals involving triginte grals or we could use the following formula.

$$\int \cos^m\theta d\theta = \frac{1}{m}\sin\theta\cos^{m-1}\theta + \frac{m-1}{m}\int \cos^{m-2}\theta d\theta$$

Let's use this formula to do the integral.

$$\int \cos^4 Od0 - \frac{1}{4}\sin\theta\cos^3\theta \mid \frac{3}{4}\int \cos^{m-2}\theta d\theta$$

$$= \frac{1}{4}\sin\theta\cos^3\theta + \frac{3}{4}\left(\frac{1}{2}\sin\theta\cos O + \frac{1}{2}\int \cos^0 Od\theta\right)\cos{}^\circ\theta = 1!$$

$$= \frac{1}{4}\sin\theta\cos^3\theta + \frac{3}{8}\sin\theta + \frac{3}{8}\sin\theta\cos\theta + \frac{3}{8}\theta$$

Next, let's use the following right triangle to get this back to x's.

$$\tan\theta = \frac{u}{\sqrt{2}} = \frac{x-3}{\sqrt{2}} \quad \sin\theta = \frac{x-3}{\sqrt{(x-3)^2+2}} \quad \cos\theta = \frac{\sqrt{2}}{\sqrt{(x-3)^2+2}}$$

$$\int \cos^4 \theta d\theta = \frac{1}{4}\frac{2\sqrt{2}(x-3)}{\left((x-3)^2+2\right)} + \frac{3}{8}\frac{\sqrt{2}(x-3)}{(x-3)^2+2} + \frac{3}{8}\tan^{-1}\left(\frac{x-3}{\sqrt{2}}\right)$$

$$= \frac{\sqrt{2}}{2}\frac{x-3}{\left((x-3)^2+2\right)^2} + \frac{3\sqrt{2}}{8}\frac{x-3}{(x-3)^2+2} + \frac{3}{8}\tan^{-1}\left(\frac{x-3}{\sqrt{2}}\right)$$

All told then the original integral is,

$$\int \frac{x}{(x^2-6x+11)^3}dx = \frac{1}{4}\frac{1}{\left((x-3)^2+2\right)^2} + \frac{3\sqrt{2}}{8}$$

$$\left(\frac{\sqrt{2}}{2}\frac{x-3}{\left((x-3)^2+2\right)^2} + \frac{3\sqrt{2}}{8}\frac{x-3}{(x-3)^2+2} + \frac{3}{8}\tan^{-1}\left(\frac{x-3}{\sqrt{2}}\right)\right)$$

$$= \frac{1}{8}\frac{3x-11}{\left((x-3)^2+2\right)^2} + \frac{9}{32}\frac{x-3}{(x-3)^2+2} + \frac{9\sqrt{2}}{64}\tan^{-1}\left(\frac{x-3}{\sqrt{2}}\right) + c$$

It's a long and messy answer, but there it is.

Example: Evaluate the following integral.

$$\int \frac{x-3}{(4-2x-x^2)^2}dx$$

Solution: As with the other problems we'll first complete the square on the denominator.

$$4-2x-x^2 = -(x^2+2x-4) = -(x^2+2x+1-1-4)$$
$$= -\left((x+1)^2-5\right) = 5-(x+1)^2$$

The integral is,

$$\int \frac{x-3}{(4-2x-x^2)^2}dx = \int \frac{x-3}{\left[5-(x+1)^2\right]^2}dx$$

Now, let's do the substitution.

$$u = x+1 \qquad x = u-1 \qquad dx = du$$

and the integral is now,

$$\int \frac{x-3}{(4-2x-x^2)^2}dx = \int \frac{u-4}{(5-u^2)^2}du$$

$$= \int \frac{u}{(5-u^2)^2}du - \int \frac{4}{(5-u^2)^2}du$$

In the first integral we'll use the substitution

$$v = 5 - u^2$$

and in the second integral we'll use the following trig substitution

$$u = \sqrt{5}\sin\theta \qquad du = \sqrt{5}\cos\theta d\theta$$

Using these substitutions the integral becomes,

$$\int \frac{x}{(4-2x+x^2)^2}dx = \frac{1}{2}\int \frac{1}{v^2}dv + \int \frac{4}{(5-5\sin^2\theta)^2}(\sqrt{5}\cos\theta)\,d\theta$$

$$= \frac{1}{2}\frac{1}{v} - \frac{4\sqrt{5}}{25}\int \frac{\cos\theta}{(1-\sin^2\theta)^2}d\theta$$

$$= \frac{1}{2}\frac{1}{v} - \frac{4\sqrt{5}}{25}\int \frac{\cos\theta}{\cos^4\theta}d\theta$$

$$= \frac{1}{2}\frac{1}{v} - \frac{4\sqrt{5}}{25}\int \sin^2\theta d\theta$$

$$= \frac{1}{2}\frac{1}{v} - \frac{2\sqrt{5}}{25}(\sec\theta\tan\theta + \ln|\sec\theta + \tan\theta|) + c$$

We'll need the following right triangle to finish this integral out.

$$\sin\theta = \frac{u}{\sqrt{5}} = \frac{x+1}{\sqrt{5}} \qquad \sec\theta = \frac{\sqrt{5}}{\sqrt{5}-(x+1)^2} \qquad \tan\theta = \frac{x+1}{\sqrt{5-}(x+1)^2}$$

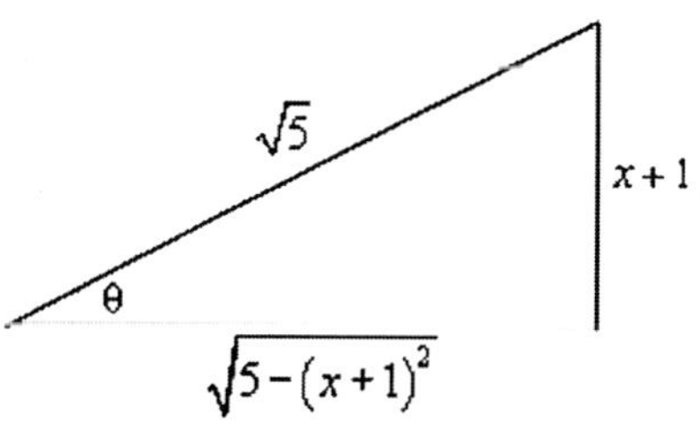

So, going back to x's the integral becomes,

$$\int \frac{x-3}{(4-2x-x^2)^2}\,dx = \frac{1}{2}\frac{1}{5-u^2}$$

$$-\frac{2\sqrt{5}}{25}\left(\frac{\sqrt{5}(x+1)}{5-(x+1)^2}\ln\left|\frac{\sqrt{5}}{\sqrt{5-(x-1)^2}}+\frac{x+1}{\sqrt{5-(x-1)^2}}\right|\right)+c$$

$$= \frac{1}{10}\frac{4x-1}{5-(x+1)^2}+\frac{2\sqrt{5}}{25}\ln\left|\frac{x+1\sqrt{5}}{\sqrt{5-(x+1)^2}}\right|+c$$

Often the following formula is needed when using the trig substitution that we used in the previous example.

$$\left|\int \sec^m \theta d\theta = \frac{1}{m-1}\tan\theta\sec^{m-2}\theta + \frac{m-2}{m-1}\int \sec^{m-2}\theta d\theta\right|$$

Note that we'll only need the two trig substitutions that we used here. The third trig substitution that we used will not be needed here.

Using Integral Tables

This section is entitled Using Integral Tables and we will be using integral tables. However, at some level, this isn't really the point of this section.

To a certain extent the real subject of this section is how to take advantage of known integrals to do integrals that may not look like anything the ones that we do know how to do or are given in a table of integrals. For the most part we'll be doing this by using substitution to put integrals into a form that we can deal with. However, not all of the integrals will require a substitution. For some integrals all that we need to do is a little rewriting of the integrand to get into a form that we can deal with. We've already related a new integral to one we could deal with least once. In the last example in the Trig Substitution section we looked at the following integral.

$$\int e^{4x}\sqrt{1+e^{2x}}\,dx$$

At first glance this looks nothing like a trig substitution problem. However, with the substitution $u = e^x$ we could turn the integral into,

$$\int u^3\sqrt{1+u^2}\,du$$

which definitely is a trig substitution problem (u = tan θ). We actually did this process in a single step by using, $e^x = \tan\theta$, but the point is that with a substitution we were able to convert an integral into a form that we could deal with.

So, let's work a couple examples using substitutions and tables.

Example: Evaluate the following integral.

$$\int \frac{\sqrt{7+9x^2}}{x^2}dx$$

Solution: So, the first thing we should do is go to the tables and see if there is anything in the tables that is close to this. In the tables in Stewart we find the following integral,

$$\int \frac{\sqrt{a^2+u^2}}{u^2}du = \frac{\sqrt{a^2+u^2}}{u} + \ln\left(u+\sqrt{a^2+u^2}\right)+c$$

This is nearly what we've got in our integral. The only real difference is that we've got a coefficient in front of the x^2 and the formula doesn't. This is easily enough dealt with. All we need to do is the following manipulation on the integrand.

$$\int \frac{\sqrt{a^2+u^2}}{u^2}dx = \int \frac{\sqrt{9\left(\frac{7}{9}+x^2\right)}}{x^2}dx = \int \frac{3\sqrt{\frac{7}{9}+x^2}}{x^2}dx = 3\int \frac{\sqrt{\frac{7}{9}+x^2}}{x^2}dx$$

So, we can now use the formula with $a = \dfrac{\sqrt{7}}{3}$.

$$\int \frac{\sqrt{7+9x^2}}{x^2}dx = 3\left(-\frac{\sqrt{\frac{7}{9}+x^2}}{x} + \ln\left(x+\sqrt{\frac{7}{9}+x^2}\right)\right)+c$$

Example: Evaluate the following integral.

$$\int \frac{\cos x}{\sin x\sqrt{9\sin x-4}}dx$$

Solution: Going through our tables we aren't going to find anything that looks like this in them.

However, notice that with the substitution we can rewrite the integral as,

$$\int \frac{\cos x}{\sin x\sqrt{9\sin x-4}}dx = \int \frac{1}{u\sqrt{9u-4}}$$

and this is in the tables.

$$\int \frac{1}{u\sqrt{a+bu}}du = \frac{1}{\sqrt{a}}\ln\left|\frac{\sqrt{a+bu}-\sqrt{a}}{\sqrt{a+bu}+\sqrt{a}}\right| + c \quad \text{if } a>0$$

$$= \frac{1}{\sqrt{-a}}\tan^{-1}\left(\sqrt{\frac{a+bu}{-a}}\right) + c \quad \text{if } a<0$$

Notice that this is a formula that will depend upon the value of a. This will happen on occasion. In our case we have a = – and b = 9 so we'll use the second formula.

$$\int \frac{\cos x}{\sin x\sqrt{9\sin x-4}}dx = \frac{2}{\sqrt{-(-4)}}\tan^{-1}\left(\sqrt{\frac{9u-4}{-(-4)}}\right) + c$$

$$= \tan^{-1}\left(\sqrt{\frac{9\sin x-4}{-a}}\right) + c$$

This final example uses a type of formula known as a reduction formula.

Example: Evaluate the following integral.

$$\int \cot^4\left(\frac{x}{2}\right)dx$$

Solution: We'll first need to use the substitution $u = \frac{x}{2}$ since none of the formulas in our tables have that in them. Doing this gives,

$$\int \cot^4\left(\frac{x}{2}\right)dx = 2\int \cot^4 u\, du$$

To help us with this integral we'll use the following formula.

$$\int \cot^n u\, du = \frac{-1}{n-1}\cot^{n-1} u - \int \cot^{n-2} u\, du$$

Formulas like this are called reduction formulas. Reduction formulas generally don't explicitly give the integral. Instead they reduce the integral to an easier one. In fact they often reduce the integral to a different version of itself!

For our integral we'll use $n = 4$.

$$\int \cot^4\left(\frac{x}{2}\right)dx = 2\left(-\frac{1}{3}\cot^3 u - \int \cot^{n-2} u\, du\right)$$

At this stage we can either reuse the reduction formula with or use the formula

$$\int \cot^2 u\, du = -\cot u - u + c$$

We'll reuse the reduction formula with $n = 2$ so we can address something that happens on occasion.

$$\int \cot^4\left(\frac{x}{2}\right)dx = 2\left(-\frac{1}{3}\cot^3 u - \left(-\frac{1}{1}\cot u - \int \cot^0 u\, du\right)\right) \cot^0 u = 1!$$

$$= -\frac{2}{3}\cot^3 u + 2\cot u + 2\int du$$

$$= -\frac{2}{3}\cot^3 u + 2\cot u + 2u + c$$

$$= -\frac{2}{3}\cot^3\left(\frac{x}{2}\right) + 2\cot\left(\frac{x}{2}\right) + x + c$$

Don't forget that $a^0 = 1$. Often people forget that and then get stuck on the final integral!

Integration Strategy

We've now seen a fair number of different integration techniques and so we should probably pause at this point and talk a little bit about a strategy to use for determining the correct technique to use when faced with an integral There are a couple of points that need to be made about this strategy.

First, it isn't a hard and fast set of rules for determining the method that should be used. It is really nothing more than a general set of guidelines that will help us to identify techniques that may work.

Some integrals can be done in more than one way and so depending on the path you take through the strategy you may end up with a different technique than somebody else who also went through this strategy.

Second, while the strategy is presented as a way to identify the technique that could be used on an integral also keep in mind that, for many integrals, it can also automatically exclude certain techniques as well. When going through the strategy keep two lists in mind. The first list is integration techniques that simply won't work and the second list is techniques that look like they might work. After going through the strategy and the second list has only one entry then that is the technique to use.

If, on the other hand, there are more than one possible technique to use we will then have to decide on which is liable to be the best for us to use. Unfortunately there is no way to teach which technique is the best as that usually depends upon the person and which technique they find to be the easiest.

Third, don't forget that many integrals can be evaluated in multiple ways and so more than one technique may be used on it. This has already been mentioned in each of the previous points, but is important enough to warrant a separate mention. Sometimes one technique will be significantly easier than the others and so don't just stop at the first technique that appears to work. Always identify all possible techniques and then go back and determine which you feel will be the easiest for you to use.

Integrand

This step is very important in the integration process. Many integrals can be taken from impossible or very difficult to very easy with a little simplification or manipulation. Don't forget basic trig and algebraic identities as these can often be used to simplify the integral. We used this idea when we were looking at integrals involving trig functions. For example consider the following integral.

$$\int \cos^2 x \, dx$$

This integral can't be done as is however, simply by recalling the identity,

$$\cos^2 x = \frac{1}{2}(1 + \cos(2x))$$

the integral becomes very easy to do.

Note that this example also shows that simplification does not necessarily mean that we'll write the integrand in a "simpler" form.

It only means that we'll write the integrand into a form that we can deal with and this is often longer and/or "messier" than the original integral.

1. See if a "simple" substitution will work. Look to see if a simple substitution can be used instead of the often more complicated methods from Calculus. For example consider both if the following integrals.

   $$\int \frac{x}{x^2 - 1} dx \qquad \int x\sqrt{x^2 - 1}dx$$

   The first integral can be done with partial fractions and the second could be done with a trig substitution.

   However, both could also be evaluated using the substitution $u = x^2 - 1$ and the work involved in the substitution would be significantly less than the work involved in either partial fractions or trig substitution.

2. Identify the type of integral. Note that any integral may fall into more than one of these types.

   Because of this fact it's usually best to go all the way through the list and identify all possible types since one may be easier than the other and it's entirely possible that the easier type is listed lower in the list.

   (a) Is the integrand a rational expression (*i.e* is the integrand a polynomial divided by a poly-nomial)? If so, then partial fractions may work on the integral.

   (b) Is the integrand a polynomial times a trig function, exponential, or logarithm? If so, then integration by parts may work.

   (c) Is the integrand a product of sines and cosines, secant and tangents, or cosecants and cotangents?

   If so, then the topics from the second section may work.

   Likewise, don't forget that some quotients involving these functions can also be done using these techniques.

(d) Does the integrand involve $\sqrt{b^2x^2 + a^2}$, $\sqrt{b^2x^2 - a^2}$, or $\sqrt{a^2 - b^2x^2}$? If so, then a trig substitution might work nicely.

(e) Does the integrand have roots other than those listed above in it? If so, then the substitution $u = \sqrt[n]{g(x)}$ might work.

(f) Does the integrand have a quadratic in it? If so, then completing the square on the quadratic might put it into a form that we can deal with.

3. Can we relate the integral to an integral we already know how to do? In other words, can we use a substitution or manipulation to write the integrand into a form that does fit into the forms we've looked at previously in this chapter.

   A typical example here is the following integral.

$$\int \cos x\sqrt{1+\sin^2 x}\, dx$$

   This integral doesn't obviously fit into any of the forms we looked at in this chapter. However, with the substitution $u = \sin x$ we can reduce the integral to the form,

$$\sqrt{1+u^2}\, du$$

   which is a trig substitution problem.

4. Do we need to use multiple techniques? In this step we need to ask ourselves if it is possible that we'll need to use multiple techniques. The example in the previous part is a good example.

   Using a substitution didn't allow us to actually do the integral.

   All it did was put the integral and put it into a form that we could use a different technique on. Don't ever get locked into the idea that an integral will only require one step to completely evaluate it. Many will require more than one step.

5. Try again. If everything that you've tried to this point doesn't work then go back through the process and try again. This time try a technique that that you didn't use the first time around.

As noted above this strategy is not a hard and fast set of rules. It is only intended to guide you through the process of best determining how to do any given integral. Steps 1, 2 and 4 involve nothing more than manipulation of the integrand either through direct manipulation of the integrand or by using a substitution.

The last two steps are simply ideas to think about in going through this strategy. Many students go through this process and concentrate almost exclusively on Step 3 to the exclusion of the other steps.

One very large consequence of that exclusion is that often a simple manipulation or substitution is overlooked that could make the integral very easy to do. Before moving on to the next section we should work a couple of quick problems illustrating a couple of not so obvious simplifications/manipulations and a not so obvious substitution.

Example: Evaluate the following integral.

$$\int \frac{\tan x}{\sec^4 x} dx$$

Solution: This integral almost falls into the form given in 3c. It is a quotient of tangent and secant and we know that sometimes we can use the same methods for products of tangents and secants on quotients.

The process from that section tells us that if we have even powers of secant to strip two of them off and convert the rest to tangents. That won't work here. We can split two secants off, but they would be in the denominator and they won't do us any good there. Remember that the point of splitting them off is so they would be there for the substitution $u = \tan x$.

That requires them to be in the numerator. So, that won't work and so we'll have to find another solution method. There are in fact two solution methods to this integral depending on how you want to go about it. We'll take a look at both.

Solution: In this solution method we could just convert everything to sines and cosines and see if that gives us an integral we can deal with.

$$\int \frac{\tan x}{\sec^4 x}\,dx = \int \frac{\sin x}{\cos x}\cos^4 x\,dx$$

$$= \int \sin x \cos^3 x\,dx \qquad u = \cos x$$

$$= -\int u^3\,du$$

$$= -\frac{1}{4}\cos^4 x + c$$

Note that just converting to sines and cosines won't always work and if it does it won't always work this nicely. Often there will be a lot more work that would need to be done to complete the integral.

Solution: This solution method goes back to dealing with secants and tangents. Let's notice that if we had a secant in the numerator we could just use $u = \sec x$ as a substitution and it would be a fairly quick and simple substitution to use. We don't have a secant in the numerator. However we could very easily get a secant in the numerator simply by multiplying the numerator and denominator by secant.

$$\int \frac{\tan x}{\sec^4 x}\,dx = \int \frac{\tan x \sec x}{\cos^5 x}\,dx \qquad u = \sec x$$

$$= \int \frac{1}{u^5}\,du$$

$$= -\frac{1}{4}\frac{1}{\sec^4 x} + c$$

$$= -\frac{1}{4}\cos^4 x + c$$

In the previous example we saw two "simplifications" that allowed us to do the integral. The first was using identities to rewrite the integral into terms we could deal with and the second involved multiplying the numerator and the denominator by something to again put the integral into terms we could deal with. Using identities to rewrite an integral is an important "simplification" and we should not forget about it.

Chaper 12

Fundamental Theorem for Line Integrals

Line Integrals Part I

In this section we are now going to introduce a new kind of integral. However, before we do that it is important to note that you will need to remember how to parameterize equations, or put another way, you will need to be able to write down a set of parametric equations for a given curve. If you need some review you should go back and review some of the basics of parametric equations and curves. Here are some of the more basic curves that we'll need to know how to do as well as limits on the parameter if they are required.

Curve	Parametric	Equations
(Ellipse)	Counter-Clockwise	Clockwise
	$x = a\cos(t)$	$x = a\cos(t)$
$\frac{x^2}{a^2} + \frac{y^2}{b^2} = 1$	$y = b\sin(t)$	$y = -b\sin(t)$
	$0 \le t \le 2\pi$	$0 \le t \le 2\pi$
(Circle)	Counter Clockwise	Clockwise
	$x = r\cos(t)$	$x = r\cos(t)$
$x^2 + y^2 = r^2$	$y = r\sin(t)$	$y = -r\sin(t)$
	$0 \le t \le 2\pi$	$0 \le t \le 2\pi$
$y = f(x)$	$x = t$	
	$y = f(t)$	

$x = g(y)$	$y = g(t)$ $y = t$
Line segment from (x_0, y_0, z_0) to (x_1, y_1, z_1)	$x = (1-t)x_0 + tx_1$ $y = (1-t)y_0 + ty_1, 0 \le t \le 1$ $z = (1-t)z_0 + tz_1$

With the final one we gave both the vector form of the equation as well as the parametric form and if we need the two-dimensional version then we just drop the z components.

In fact, we will be using the two-dimensional version of this in this section.

For the ellipse and the circle we've given two parameterizations, one tracing out the curve clockwise and the other counter-clockwise. As we'll eventually see the direction that the curve is traced out can, on occasion, change the answer. Also, both of these "start" on the positive x-axis at $t = 0$.

Now let's move on to line integrals $f(x)$. We integrated, a function of a single variable, over an interval $[a,b]$. In this case we were thinking of x as taking all the values in this interval starting at a and ending at b. With line integrals we will start with integrating the function $f(x,y)$, a function of two variables, and the values of x and y that we're going to use will be the points, (x,y), that lie on a curve C.

Note that this is different from the double integrals that we were working with in the previous chapter where the points came out of some two-dimensional region.

Let's start with the curve C that the points come from. We will assume that the curve is *smooth* (defined shortly) and is given by the parametric equations,

$$x = h(t) \quad y = g(t) \quad a \le t \le b$$

We will often want to write the parameterization of the curve as a vector function. In this case the curve is given by,

$$\vec{r}(t) = h(t)\vec{i} + g(t)\vec{j} \quad a \le t \le b$$

The curve is called smooth if $\vec{r}(t)$ is continuous and $\vec{r}(t) \neq 0$ for all t. The line integral of $f(x, y)$ along C is denoted by,

$$\int_C f(x, y)\, ds$$

We use a *ds* here to acknowledge the fact that we are moving along the curve, *C*, instead of the *x*-axis (denoted by *dx*) or the *y*-axis (denoted by *dy*).

Because of the *ds* this is sometimes called the line integral of *f* with respect to arc length. We've seen the notation *ds* before. When we looked at the arc length of a curve given by parametric equations we found it to be,

$$L = \int_a^b ds, \qquad where\ ds = \sqrt{\left(\frac{dx}{dt}\right)^2 + \left(\frac{dy}{dt}\right)^2}\, dt$$

It is no coincidence that we use *ds* for both of these problems. The *ds* is the same for both the arc length integral and the notation for the line integral. So, to compute a line integral we will convert everything over to the parametric equations. The line integral is then,

$$\int_C f(x, y)\, ds = \int_a^b f(h(t), g(t)) \sqrt{\left(\frac{dx}{dt}\right)^2 + \left(\frac{dy}{dt}\right)^2}\, dt$$

Don't forget to plug the parametric equations into the function as well. If we use the vector form of the parameterization we can simplify the notation up somewhat by noticing that,

$$\sqrt{\left(\frac{dx}{dt}\right)^2 + \left(\frac{dy}{dt}\right)^2} = \left\|\vec{r}\,'(t)\right\|$$

where $\left\|\vec{r}(t)\right\|$ is the magnitude or norm of $\vec{r}(t)$. Using this notation the line integral becomes,

$$\int_C f(x,y)\,ds = \int_a^b f\left(h(t),g(t)\right)\left\|\vec{r}\,'(t)\right\|dt$$

Note that as long as the parameterization of the curve C is traced out exactly once at t increases from a to b the value of the line integral will be independent of the parameterization of the curve.

Example: Evaluate $\int_C xy^4\,ds$ where C is the right half of the circle, $x^2 + y^2 = 16$. rotated in the counter clockwise direction.

Solution: We first need a parameterization of the circle. This is given by,

$$x = 4\cos t$$

$$y = 4\sin t$$

We now need a range of t's that will give the right half of the circle. The following range of t's will do this.

$$-\frac{\pi}{2} \le t \le \frac{\pi}{2}$$

Now, we need the derivatives of the parametric equations and let's compute ds.

$$\frac{dx}{dt} = -4\sin t$$

$$\frac{dy}{dt} = 4\cos t$$

$$ds = \sqrt{16\sin^2 t + 16\cos^2 t}\,dt = 4dt$$

Next we need to talk about line integrals over piecewise smooth curves.

A piecewise smooth curve is any curve that can be written as the union of a finite number of smooth curves, $C_1,\ldots,C_n$ where the end point of is the starting point of C_t. Below is an illustration of a piecewise smooth curve.

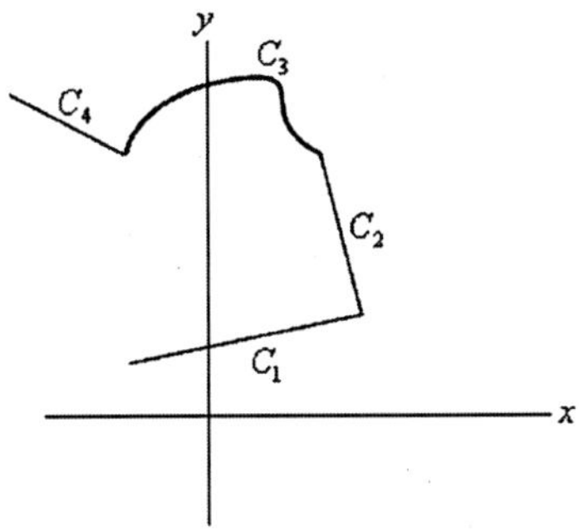

Evaluation of line integrals over piecewise smooth curves is a relatively simple thing to do. All we do is evaluate the line integral over each of the pieces and then add them up. The line integral for some function over the above piecewise curve would be,

$$\int_C f(x,y)\,ds = \int_{C_1} f(x,y)\,ds + \int_{C_2} f(x,y)\,ds + \int_{C_3} f(x,y)\,ds + \int_{C_4} f(x,y)\,ds$$

Let's see an example of this.

Example: Evaluate were $\int_C 4x^3\,ds$ is the line segment from $(-2,-1)$ to $(2,1)$.

Solution: From the parameterization formulas at the start of this section we know that the line segment start at $(-2,-1)$ and ending at $(2,1)$ is given by,

$$\begin{aligned}\vec{r}(t) &= (1-t)\langle -2,-1\rangle + t\langle 1,2\rangle \\ &= \langle -2+3t,-1+3t\rangle\end{aligned}$$

for $0 \le t \le 1$. This means that the individual parametric equations are,

$$x = -2+3t$$
$$y = -1+3t$$

Using this path the line integral is,

$$\int_C 4x^3\,ds = \int_0^1 4(-2+3t)^3\sqrt{9+9}\,dt$$

$$= 12\sqrt{2}\left(\frac{1}{12}\right)(-2+3t)^4\Big|_0^1$$

$$= 12\sqrt{2}\left(-\frac{5}{4}\right)$$

$$= -15\sqrt{2} = -21.213$$

When doing these integrals don't forget simple substitutions to avoid having to do things like cubing out that term. Cubing it out is not that difficult, but is more work than a simple substitution.

So, the previous two examples seem to suggest that if we change the path between two points then the value of the line integral (with respect to arc length) will change.

While this will happen fairly regularly we can't assume that it will always happen. In a later section we will investigate this idea in more detail

Example: Evaluate $\int_C 4x^3\,ds$ were C is the line segment from $(-2,-1)$ to $(2,1)$.

Solution: This one isn't much different, work wise, from the previous example. Here is the parameterization of the curve.

$$\vec{r}(t) = (1-t)\langle 1,2\rangle + t\langle -2,-1\rangle$$

$$= \langle 1-3t, 2-3t\rangle$$

for $0 \le t \le 1$. Remember that we are switch the direction of the curve and this will also change the parameterization so we can make sure that we start/end at the proper point.

Here is the line integral.

$$\int_C 4x^3\,ds = \int_0^1 4(1-3t)^3\sqrt{9+9}$$

$$= 12\sqrt{2}\left(\frac{1}{12}\right)(-2+3t)^4\Big|_0^1$$

$$= 12\sqrt{2}\left(-\frac{5}{4}\right)$$

$$= -15\sqrt{2} = -21.213$$

$$= 12\sqrt{2}\left(\frac{1}{12}\right)(-2+3t)^4\Big|_0^1$$

$$= 12\sqrt{2}\left(-\frac{5}{4}\right)$$

$$= -15\sqrt{2} = -21.213$$

So, it looks like when we switch the direction of the curve the line integral (with respect to arc length) will not change. This will always be true for these kinds of line integrals.

However, there are other kinds of line integrals in which this won't be the case.

We will see more examples of this in the next couple of sections so don't get it into your head that changing the direction will never change the value of the line integral.

Before working another example let's formalize this idea up somewhat. Let's suppose that the curve C has the parameterization $x = h(t), y = g(t)$. Let's also suppose that the initial point on the curve is A and the final point on the curve is B.

The parameterization $x = h(t), y - g(t)$ will then determine an orientation for the curve where the positive direction is the direction that is traced out as t increases.

Finally, let $-C$be the curve with the same points as C, however in this case the curve has B as the initial point and A as the final point, again t is increasing as we traverse this curve.

In other words, given a curve C, the curve $-C$is the same curve as C except the direction has been reversed.

We then have the following fact about line integrals with respect to arc length.

Fact

$$\int\limits_{C} f(x,y)\,ds = \int\limits_{-C} f(x,y)\,ds$$

So, for a line integral with respect to arc length we can change the direction of the curve and not change the value of the integral.

This is a useful fact to remember as some line integrals will be easier in one direction than the other.

Example: Evaluate for each of the following curves.

1. $C_1 : y = x^2, -1 \le x \le 1$
2. C_2 : The line segment from $(-1,1)$ to $(1,1)$.
3. C_3 : The line segment from $(1,1)$ to $(-1,1)$.

Solution: Before working any of these line integrals let's notice that all of these curves are paths that connect the points $(-1,1)$ and $(1,1)$. Also notice that $C_3 = -C_2$ and so by the fact above these two should give the same answer.

Here is a sketch of the three curves and note that the curves illustrating and have been separated a little to show that they are separate curves in some way even thought they are the same line.

1. $C_1 : y = x^2, -1 \le x \le 1$

Here is a parameterization for this curve.

$$C_1 : x = t, y = t^2, -1 \le t \le 1$$

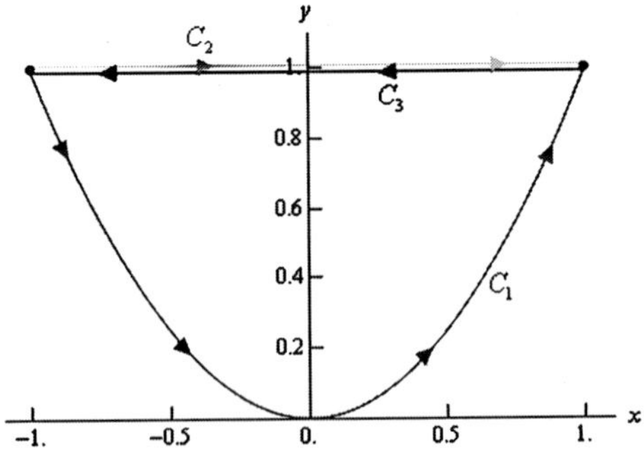

Here is the line integral.

$$\int_{C} x\,ds = \int_{-1}^{1} t\sqrt{1+4t^2}\,dt = \frac{1}{12}\left(1+4t^2\right)^{\frac{3}{2}}\Bigg|_{-1}^{1} = 0$$

2. C_2 : The line segment from $(-1,1)$ to $(1,1)$.

There are two parameterizations that we could use here for this curve. The first is to use the formula we used in the previous couple of examples. That parameterization is,

$$\begin{aligned} C_2 : \vec{r}(t) &= (1-t)\langle -1,1\rangle + t\langle 1,1\rangle \\ &= \langle 2t-1,1\rangle \end{aligned}$$

for $0 \le t \le 1$.

Sometimes we have no choice but to use this parameterization. However, in this case there is a second (probably) easier parameterization.

The second one uses the fact that we are really just graphing a portion of the line $y = 1$. Using this the parameterization is,

$$C_2 : x = t, y = 1, -1 \le t \le 1$$

This will be a much easier parameterization to use so we will use this. Here is the line integral for this curve.

$$\int_{C} x\,ds = \int_{-1}^{1} t\sqrt{1+0}\,dt = \frac{1}{2}t^2\Bigg|_{-1}^{1} = 0$$

Note that this time, unlike the line integral we worked with in Examples we got the same value for the integral despite the fact that the path is different.

This will happen on occasion. We should also not expect this integral to be the same for all paths between these two points.

At this point all we know is that for these two paths the line integral will have the same value.

It is completely possible that there is another path between these two points that will give a different value for the line integral.

3. C_3: The line segment from $(1,1)$ to $(-1,1)$.

Now, according to our fact above we really don't need to do anything here since we know that $C_3 = -C_2$. The fact tells us that this line integral should be the same as the second part (*i.e.* zero).

However, let's verify that, plus there is a point we need to make here about the parameterization.

Here is the parameterization for this curve.

$$\begin{aligned} C_3 : \vec{r}(t) &= (1-t)\langle 1,1\rangle + t\langle -1,1\rangle \\ &= \langle 1-2t, 1\rangle \end{aligned}$$

for $0 \le t \le 1$.

Note that this time we can't use the second parameterization that we used in part (b) since we need to move from right to left as the parameter increases and the second parameterization used in the previous part will move in the opposite direction.

Here is the line integral for this curve.

$$\int_{C_3} x\,ds = \int_{-1}^{1} (1-2t)\sqrt{4+0}\,dt = 2\left(t-t^2\right)\Big|_0^1 = 0$$

Sure enough we got the same answer as the second part.

To this point in this section we've only looked at line integrals over a two-dimensional curve.

However, there is no reason to restrict ourselves like that. We can do line integrals over three-dimensional curves as well. Let's suppose that the three-dimensional curve C is given by the parameterization,

$$\begin{aligned} x &= x(t) \\ y &= y(t) \\ z &= z(t) \\ z \le\ & t \le b \end{aligned}$$

then the line integral is given by,

$$\int_C f(x,y,z)ds = \int_a^b f(x(t),y(t),z(t))$$

$$\sqrt{\left(\frac{dx}{dt}\right)^2+\left(\frac{dy}{dt}\right)^2+\left(\frac{dz}{dt}\right)^2}\,dt$$

Note that often when dealing with three-dimensional space the parameterization will be given as a vector function.

$$\vec{r}(t) = \langle x(t), y(t), z(t)\rangle$$

Notice that we changed up the notation for the parameterization a little. Since we rarely use the function names we simply kept the x, y, and z and added on the (t)part to denote that they may be functions of the parameter. Also notice that, as with two-dimensional curves, we have,

$$\sqrt{\left(\frac{dx}{dt}\right)^2+\left(\frac{dy}{dt}\right)^2+\left(\frac{dz}{dt}\right)^2} = \|\vec{r}'(t)\|$$

and the line integral can again be written as,

$$\int_C f(x,y,z)\,ds = \int_a^b f(x(t),y(t),z(t))\|\vec{r}'(t)\|dt$$

So, outside of the addition of a third parametric equation line integrals in three-dimensional space work the same as those in two-dimensional space. Let's work a quick example.

Example: Evaluate $\int_C xyz\,ds$ where C is the helix given by $\|\vec{r}(t)\| = \langle \cos(t), \sin(t), 3t\rangle$, $0 \le t \le 4\pi$.

Solution: Note that we first saw the vector equation for a helix back in the Vector Functions section. Here is a quick sketch of the helix. Here is the line integral.

$$\int_C xyz\,ds = \int_0^{4\pi} 3t\cos(t)\sin(t)\sqrt{\sin^2 t+\cos^2 t+9}dt$$

$$= \int_0^{4\pi} 3t\left(\frac{1}{2}\sin(2t)\right)\sqrt{1+9}$$

$$= \frac{3\sqrt{10}}{2}\left(\frac{1}{4}\sin(2t) - \frac{t}{2}\cos(2t)\right)\Bigg|_0^{4\pi}$$

$$= -3\sqrt{10\pi}$$

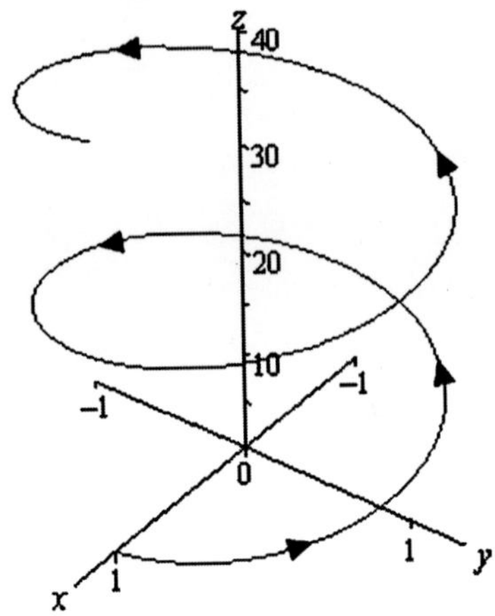

You were able to do that integral right? It required integration by parts.

Chapter 13

Methods of Integration

Integrals Involving Trig Functions

In this section we are going to look at quite a few integrals involving trig functions and some of the techniques we can use to help us evaluate them. Let's start off with an integral that we should already be able to do.

$$\int \cos x \sin^5 x\,dx = \int u^5\,du \quad \text{using the substitution } u = \sin x$$
$$= \frac{1}{6}\sin^6 x + c$$

This integral is easy to do with a substitution because the presence of the cosine, however, what about the following integral.

Example: Evaluate the following integral.

$$\int \sin^5 x dx$$

Solution: This integral no longer has the cosine in it that would allow us to use the substitution that we used above. Therefore, that substitution won't work and we are going to have to find another way of doing this integral.

Let's first notice that we could write the integral as follows,

$$\int \sin^5 x dx = \int \sin^4 x \sin x dx = \int \left(\sin^2 x\right)^2 \sin x dx$$

Now recall the trig identity,

$$\cos^2 x + \sin^2 x + 1 \qquad \Rightarrow \qquad \sin^2 x + 1 - \cos^2 x$$

With this identity the integral can be written as,

$$\int \sin^5 x dx = \int \left(1 - \cos^2 x\right)^2 \sin x dx$$

and we can now use the substitution $u = cos\ x$. Doing this gives us,

$$\begin{aligned}\int \sin^5 x\, dx &= \int (1 - u^2)^2 du \\ &= -\int 1 - 2u^2 + u^4\, du \\ &= -\left(u - \frac{2}{3}u^3 + \frac{1}{5}u^5\right) + c \\ &= -\cos x + \frac{2}{3}\cos^3 x - \frac{1}{5}\cos^5 x + c\end{aligned}$$

So, with a little rewriting on the integrand we were able to reduce this to a fairly simple substitution.

Notice that we were able to do the rewrite that we did in the previous example because the exponent on the sine was odd. In these cases all that we need to do is strip out one of the sines. The exponent on the remaining sines will then be even and we can easily convert the remaining sines to cosines using the identity,

$$\cos^2 x + \sin^2 x = 1$$

If the exponent on the sines had been even this would have been difficult to do.

We could strip out a sine, but the remaining sines would then have an odd exponent and while we could convert them to cosines the resulting integral would often be even more difficult than the original integral in most cases.

Let's take a look at another example.

Example: Evaluate the following integral.

$$\int \sin^6 x \cos^3 x dx$$

Solution: So, in this case we've got both sines and cosines in the problem and in this case the exponent on the sine is even while the exponent on the cosine is odd. So, we can use a similar technique in this integral. This time we'll strip out a cosine and convert the rest to sines.

$$\int \sin^6 x \cos^3 x\, dx = \int \sin^6 x \cos^2 x \cos x\, dx$$
$$= \int \sin^6 x(1-\sin^2 x)\cos x\, dx \qquad u = \sin x$$
$$= -\left(u - \frac{2}{3}u^3 + \frac{1}{5}u^5\right) + c$$
$$= \int u^6 (1-u^2)\, dx$$
$$= \int u^6 - u^8 du$$
$$= \frac{1}{7}\sin^7 x - \frac{1}{9}\sin^9 x + c$$

At this point let's pause for a second to summarize what we've learned so far about integrating powers of sine and cosine.

$$\int \sin^n x \cos^m x dx$$

In this integral if the exponent on the sines (n) is odd we can strip out one sine, convert the rest to cosines using and then use the substitution. Likewise, if the exponent on the cosines (m) is odd we can strip out one cosine and convert the rest to sines and the use the substitution $u = sin\ x$. Of, course if both exponents are odd then we can use either method. However, in these cases it's usually easier to convert the term with the smaller exponent. The one case we haven't looked at is what happens if both of the exponents are even? In this case the technique we used in the first couple of examples simply won't work and in fact there really isn't any one set method for doing these integrals. Each integral is different and in some cases there will be more than one way to do the integral. With that being said most, if not all, of integrals involving products of sines and cosines in which both exponents are even can be done using one or more of the following formulas to rewrite the integrand.

$$\sin^2 x = \frac{1}{2}(1 + \sin(2x))$$
$$\cos^2 x = \frac{1}{2}(1 + \cos(2x))$$
$$\sin x \cos x = \frac{1}{2}\sin(2x))$$

The first two formulas are the standard half angle formula from a trig class written in a form that will be more convenient for us to use. The last is the standard double angle formula for sine, again with a small rewrite.

Let's take a look at an example.

Example: Evaluate the following integral.

$$\int \sin^2 \cos^2 x dx$$

Solution: As noted above there are often more than one way to do integrals in which both of the exponents are even. This integral is an example of that. There are at least two solution techniques for this problem. We will do both solutions starting with what is probably the harder of the two, but it's also the one that many people see first. In this solution we will use the two half angle formulas above and just substitute them into the integral.

$$\int \sin^2 \cos^2 x dx = \int \frac{1}{2}(1-\cos(2x))\left(\frac{1}{2}\right)(1+\cos(2x))dx$$

$$= \frac{1}{4}\int 1-\cos^2(2x)dx$$

So, we still have an integral that can't be completely done, however notice that we have managed to reduce the integral down to just one term causing problems (a cosine with an even power) rather than two terms causing problems. In fact to eliminate the remaining problem term all that we need to do is reuse the first half angle formula given above.

$$\int \sin^2 x \cos^2 x\, dx = \frac{1}{4}\int 1-\frac{1}{2}(1+\cos(4x))\, dx$$

$$= \frac{1}{4}\int \frac{1}{2}-\frac{1}{2}\cos(4x))\, dx$$

$$= \frac{1}{4}\left(\frac{1}{2}x-\frac{1}{8}\sin(4x)\right)+c$$

$$= \frac{1}{8}x-\frac{1}{32}\sin(4x)+c$$

So, this solution required a total of three trig identities to complete.

Solution: In this solution we will use the half angle formula to help simplify the integral as follows.

$$\begin{aligned}\int \sin^2 x \cos^2 x\,dx &= \int (\sin x \cos x)^2\,dx \\ &= \int \left(\frac{1}{2}\sin(2x)\right)^2 dx \\ &= \frac{1}{4}\int \sin^2(2x)\,dx\end{aligned}$$

Now, we use the double angle formula for sine to reduce to an integral that we can do.

$$\begin{aligned}\int \sin^2 x \cos^2 x\,dx &= \frac{1}{8}\int 1-\cos(4x)\,dx \\ &= \frac{1}{8}x + \frac{1}{32}\sin(4x) + c\end{aligned}$$

This method required only two trig identities to complete. Notice that the difference between these two methods is more one of "messiness".

The second method is not appreciably easier (other than needing one less trig identity) it is just not as messy and that will often translate into an "easier" process. In the previous example we saw two different solution methods that gave the same answer. Note that this will not always happen. In fact, more often than not we will get different answers. However, as we discussed in the Integration by Parts section, the two answers will differ by no more than a constant.

In general when we have products of sines and cosines in which both exponents are even we will need to use a series of half angle and/or double angle formulas to reduce the integral into a form that we can integrate. Also, the larger the exponents the more we'll need to use these formulas and hence the messier the problem. Sometimes in the process of reducing integrals in which both exponents are even we will run across products of sine and cosine in which the arguments are different.

These will require one of the following formulas to reduce the products to integrals that we can do.

$$\sin\alpha\cos\beta = \frac{1}{2}\left[\sin(\alpha+\beta)+\sin(\alpha+\beta)\right]$$

$$\sin\alpha\sin\beta = \frac{1}{2}\left[\sin(\alpha-\beta)-\cos(\alpha+\beta)\right]$$

$$\cos\alpha\cos\beta = \frac{1}{2}\left[\cos(\alpha-\beta)+\cos(\alpha-\beta)\right]$$

Let's take a look at an example of one of these kinds of integrals.

Example: Evaluate the following integral.

$$\int \cos(15x)\cos(4x)\,dx$$

Solution: This integral requires the last formula listed above.

$$\begin{aligned}\int \cos(15x)\cos(4x)\,dx &= \frac{1}{2}\int \cos(11x)+\cos(19x)\,dx \\ &= \frac{1}{2}\left(\frac{1}{11}\sin(11x)+\frac{1}{19}\sin(19x)\right)+c\end{aligned}$$

Okay, at this point we've covered pretty much all the possible cases involving products of sines and cosines.

It's now time to look at integrals that involve products of secants and tangents.

This time, let's do a little analysis of the possibilities before we just jump into examples. The general integral will be,

$$\int \sec^n x \tan^m x dx$$

The first thing to notice is that we can easily convert even powers of secants to tangents and even powers of tangents to secants by using a formula.

In fact, the formula can be derived so let's do that.

$$\sin^2 x + \cos^2 x = 1$$

$$\frac{\sin^2 x}{\cos^2 x} + \frac{\cos^2 x}{\cos^2 x} = \frac{1}{\cos^2 x} \qquad \tan^2 x + 1 = \sec^2 x$$

We'll want to eventually use one of the following substitutions.

$$u = \tan x \qquad du = \sec^2 x dx$$

$$u = \sec x \qquad du = \sec x \tan x dx$$

So, if we use the substitution we will need two secants left for the substitution to work. This means that if the exponent on the secant (n) is even we can strip two out and then convert the remaining secants to tangents. Next, if we want to use the substitution $u = sec\ x$ we will need one secant and one tangent left over in order to use the substitution. This means that if the exponent on the tangent (m) is odd and we have at least one secant in the integrand we can strip out one of the tangents along with one of the secants of course. The tangent will then have an even exponent to convert the rest to tangents to secants.

Note that this method does require that we have at least one secant in the integral as well. If there aren't any secants then we'll need to do something different. If the exponent on the secant is even and the exponent on the tangent is odd then we can use either case. Again, it will be easier to convert the term with the smallest exponent.

Let's take a look at a couple of examples.

Example: Evaluate the following integral.

$$\int \sec^9 x \tan^5 x dx$$

Solution: First note that since the exponent on the secant isn't even we can't use the substitution. However, the exponent on the tangent is odd and we've got a secant in the integral and so we will be able to use the substitution $u = sec\ x$.

This means striping out a single tangent (along with a secant) and converting the remaining tangents to secants. Here's the work for this integral.

$$\begin{aligned}\int \sec^9 x \tan^5 x\, dx &= \int \sec^8 x \tan^4 x \tan x \sec x\, dx \\ &= \int \sec^8 x(\sec^2 x - 1)^2 \tan x \sec x\, dx \qquad u = \sec x \\ &= \int u^8 (u^2 - 1)^2 du \\ &= \int u^{12} - 2u^{10} + u^8 du\end{aligned}$$

Example: Evaluate the following integral.

$$\int \sec^4 x \tan^6 x dx$$

Solution: So, in this example the exponent on the tangent is even so the substitution won't work.

The exponent on the secant is even and so we can use the substitution for this integral.

That means that we need to strip out two secants and convert the rest to tangents.

Here is the work for this integral.

$$\begin{aligned}\int \sec^4 x \tan^6 x\, dx &= \int \sec^2 x \tan^6 x \sec^2 x\, dx \\ &= \int (\tan^2 x + 1) \tan^6 x \sec^2 x\, dx \qquad u = \tan x \\ &= \int (u^2 - 1)^2 u^6 du \\ &= \int u^8 + u^6\, du \\ &= \frac{1}{9}\tan^9 x + \frac{1}{7}\tan^7 x + c\end{aligned}$$

Both of the previous examples fit very nicely into the patterns discussed above and so were not all that difficult to work.

However, there are a couple of exceptions to the patterns above and in these cases there is no single method that will work for every problem. Each integral will be different and may require different solution methods in order to evaluate the integral. Let's first take a look at a couple of integrals that have odd exponents on the tangents, but no secants. In these cases we can't use the substitution $u = sec\ x$ since it requires there to be at least one secant in the integral.

Example: Evaluate the following integral.

$$\int \tan x\, dx$$

Solution: To do this integral all we need to do is recall the definition of tangent in terms of sine and cosine and then this integral is nothing more than a Calculus substitution.

$$\int \tan x\,dx = \int \frac{\sin x}{\cos x}dx = \cos x$$
$$= -\int \frac{1}{u}du$$
$$= -\ln\left|\cos x\right| + c \qquad r\ln x = \ln x^{r}$$
$$= \ln\left|\cos x\right|^{1} + c$$
$$= \ln\left|\sec x\right| + c$$

Example: Evaluate the following integral.

$$\int \tan^3 xdx$$

Solution: The trick to this one is do the following manipulation of the integrand.

$$\int \tan^3 xdx = \int \tan x \tan^2 xdx$$
$$= \int \tan x\left(\sec^2 x + 1\right)$$
$$= \int \tan x \sec^2 xdx - \int \tan xdx$$

We can now use the substitution on the first integral and the results from the previous example to on the second integral.

The integral is then, $\int \tan^3 xdx = \frac{1}{2}\tan^2 x - In\left|\sec x\right| + c$

Note that all odd powers of tangent (with the exception of the first power) can be integrated using the same method we used in the previous example. For instance,

$$\int \tan^2 xdx = \int \tan^3 x\left(\sec^2 x \quad 1\right)dx = \int \tan^3 x \sec^2 xdx - \int \tan^3 xdx$$

So, a quick substitution (u = *tan x*) will give us the first integral and the second integral will always be the previous odd power. Now let's take a look at a couple of examples in which the exponent on the secant is odd and the exponent on the tangent is even. In these cases the substitutions used above won't work. It should also be noted that both of the following two integrals are integrals that we'll be seeing on occasion in later sections of this chapter and in later chapters. Because of this it wouldn't be a bad idea to make a note of these results so you'll have them ready when you need them later.

Example: Evaluate the following integral.

$$\int \sec xdx$$

Solution: This one isn't too bad once you see what you've got to do. By itself the integral can't be done. However, if we manipulate the integrand as follows we can do it.

$$\int \sec x\,dx = \int \frac{\sec x(\sec x + \tan x)}{\sec x + \tan x}dx$$
$$= \int \frac{\sec^2 x \tan x \sec x}{\sec x + \tan x}dx$$

In this form we can do the integral using the substitution $u = sec\ x + tan\ x$. Doing this gives,

$$\int \sec xdx = \text{In}\left|\text{sec x tan x}\right| + c$$

The idea used in the above example is a nice idea to keep in mind. Multiplying the numerator and denominator of a term by the same term above can, on occasion, put the integral into a form that can be integrated. Note that this method won't always work and even when it does it won't always be clear what you need to multiply the numerator and denominator by. However, when it does work and you can figure out what term you need it can greatly simplify the integral.

Here's the next example.

Example: Evaluate the following integral.

$$\int \sec^3 xdx$$

Solution: This one is different from any of the other integrals that we've done in this section. The first step to doing this integral is to perform integration by parts using the following choices for u and dv.

$$u = \sec x \qquad dv = \sec^2 xdx$$
$$du = \sec x \tan xdx \qquad v = \tan x$$

Note that using integration by parts on this problem is not an obvious choice, but it does work very nicely here. After doing integration by parts we have,

$$\int \sec^2 xdx = \sec x \tan x - \int \sec x \tan^2 xdx$$

Now the new integral also has an odd exponent on the secant and an even exponent on the tangent and so the previous examples of products of secants and tangents still won't do us any good. To do this integral we'll first write the tangents in the integral in terms of secants.

Again, this is not necessarily an obvious choice but it's what we need to do in this case.

$$\int \sec^2 x dx = \sec x \tan x - \int \sec x\left(\sec^2 x - 1\right)dx$$

$$= \sec x \tan x - \int \sec^3 x dx + \int \sec x dx$$

Now, we can use the results from the previous example to do the second integral and notice that the first integral is exactly the integral we're being asked to evaluate with a minus sign in front. So, add it to both sides to get,

$$2\int \sec^2 x dx = \sec x \tan x + \text{In}\left|\sec x + \tan x\right|$$

Finally divide by a two and we're done.

$$\int \sec^3 x dx = \frac{1}{2}\left(\sec x \tan x + \text{In}\left|\sec x + \tan x\right|\right) + c$$

Again, note that we've again used the idea of integrating the right side until the original integral shows up and then moving this to the left side and dividing by its coefficient to complete the evaluation.

We first saw this in the Integration by Parts section and noted at the time that this was a nice technique to remember. Here is another example of this technique. Now that we've looked at products of secants and tangents let's also acknowledge that because we can relate cosecants and cotangents by

$$1 + \cot^2 x = \csc^2 x$$

all of the work that we did for products of secants and tangents will also work for products of cosecants and cotangents. We'll leave it to you to verify that. There is one final topic to be discussed in this section before moving on. To this point we've looked only at products of sines and cosines and products of secants and tangents. However, the methods used to do these integrals can also be used on some quotients involving sines and cosines and quotients involving secants and tangents (and hence quotients involving cosecants and cotangents). Let's take a quick look at an example of this.

Example: Evaluate the following integral.

$$\int \frac{\sin^7 x}{\cos^4 x} dx$$

Solution: If this were a product of sines and cosines we would know what to do. We would strip out a sine (since the exponent on the sine is odd) and convert the rest of the sines to cosines.

The same idea will work in this case. We'll strip out a sine from the numerator and convert the rest to cosines as follows,

$$\int \frac{\sin^7 x}{\cos^4 x} dx = \int \frac{\sin^6 x}{\cos^4 x} \sin x \, dx$$

$$= \int \frac{(\sin^2 x)^3}{\cos^4 x} \sin x \, dx$$

$$= \int \frac{(1-\cos^2 x)^3}{\cos^4 x} \sin x \, dx$$

At this point all we need to do is use the substitution and we're done.

$$\int \frac{\sin^7 x}{\cos^4 x} dx = \int \frac{(1-u^2)^3}{u^4} du$$

$$= -\int u^{-4} - 3u^{-2} + 3 - u^2 du$$

$$= -\left(-\frac{1}{3}\frac{1}{u^3} + 3\frac{1}{u} + 3u - \frac{1}{3}u^3 \right) + c$$

$$= \frac{1}{3\cos^3 x} - \frac{3}{\cos x} - 3\cos x + \frac{1}{3}\cos^3 x + c$$

Trig Substitutions

As we have done in the last couple of sections, let's start off with a couple of integrals that we should already be able to do with a standard substitution.

$$\int x\sqrt{25x^2-4}\,dx = \frac{1}{75}(25x^2-4)^{\frac{3}{2}} + c$$

$$\int \frac{x}{\sqrt{25x^2-4}}\,dx = \frac{1}{25}\sqrt{25x^2-4} + c$$

Both of these used the substitution $u = 25x^2 - 4$ and at this point should be pretty easy for you to do. However, let's take a look at the following integral.

Example: Evaluate the following integral.

$$\int \frac{\sqrt{25x^2-4}}{x}\,dx$$

Solution: In this case the substitution $u = 25x^2 - 4$ will not work and so we're going to have to do something different for this integral. It would be nice if we could get rid of the square root somehow. The following substitution will do that for us.

$$x = \frac{2}{5}\sec\theta$$

Do not worry about where this came from at this point. As we work the problem you will see that it works and that if we have a similar type of square root in the problem we can always use a similar substitution. Before we actually do the substitution however let's verify the claim that this will allow us to get rid of the square root.

$$\sqrt{25x^2-4} = \sqrt{25\left(\frac{4}{25}\right)\sec^2\theta - 4} = \sqrt{4(\sec^2\theta - 1)} = 2\sqrt{\sec^2\theta - 1}$$

To get rid of the square root all we need to do is recall the relationship,

$$\tan^2\theta + 1 = \sec^2\theta \qquad \Rightarrow \qquad \sec^2\theta - 1 = \tan^2\theta$$

Using this fact the square root becomes,

$$\sqrt{25x^2-4} = 2\sqrt{\tan^2\theta} = 2\left|\tan\theta\right|$$

Note the presence of the absolute value bars there. These are important. Recall that

$$\sqrt{x^2} = |x|$$

There should always be absolute value bars at this stage. If we knew that $tan\theta$ was always positive or always negative we could eliminate the absolute value bars using,

$$|x| = \begin{cases} x & \text{if } x \geq 0 \\ -x & \text{if } x \geq 0 \end{cases}$$

Without limits we won't be able to determine if $tan\ \theta$ is positive of negative, however, we will need to eliminate them in order to do the integral. Therefore, since we doing an indefinite integral we will assume that $tan\ \theta$ will be positive and so we can drop the absolute value bars. This gives, $\sqrt{25x^2 - 4} = 2\tan\theta$

So, we were able to eliminate the square root using this substitution. Let's now do the substitution and see what we get. In doing the substitution don't forget that well also need to substitute for the *dx*. This is easy enough to get from the substitution.

$$x = \frac{2}{5}\sec\theta \qquad \Rightarrow \qquad dx = \frac{2}{5}\sec\theta\tan\theta d\theta$$

Using this substitution the integral becomes,

$$\int \frac{\sqrt{25x^2 - 4}}{x} dx = \int \frac{2\tan\theta}{\frac{2}{5}\sec\theta}\left(\frac{2}{5}\sec\theta\tan\theta\right)d\theta$$

$$= 2\int \tan^2\theta d\theta$$

With this substitution we were able to reduce the given integral to an integral involving trig functions and we saw how to do these problems in the previous section.

Let's finish the integral.

$$\int \frac{\sqrt{25x^2-4}}{x}\,dx = 2\int \sec^2\theta - 1\,d\theta$$
$$= 2(\tan\theta - \theta) + c$$

So, we've got an answer for the integral. Unfortunately the answer isn't given in x's as it should be. So, we need to write our answer in terms of x. We can do this with some right triangle trig. From our original substitution we have,

$$\sec\theta = \frac{5x}{2} = \frac{\text{hypotenuse}}{\text{adjacent}}$$

This gives the following right triangle.

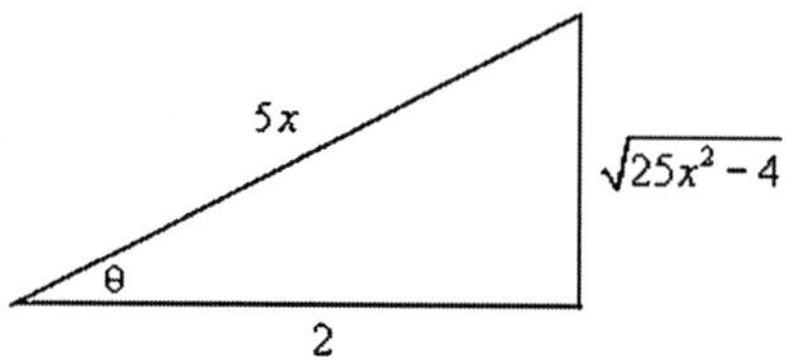

From this we can see that,

$$\tan\theta = \frac{\sqrt{25x^2-4}}{2}$$

We can deal with the in one of any variety of ways. From our substitution we can see that,

$$\theta = \text{scc}^{-1}\left(\frac{5x}{2}\right)$$

While this is a perfectly acceptable method of dealing with the we can use any of the possible six inverse trig functions and since sine and cosine are the two trig functions most people are familiar with we will usually use the inverse sine or inverse cosine. In this case we'll use the inverse cosine.

$$\theta = \cos^{-1}\left(\frac{2}{5x}\right)$$

So, with all of this the integral becomes,

$$\int \frac{\sqrt{25x^2-4}}{x}dx = 2\left(\frac{\sqrt{25x^2-4}}{x} - \cos^{-2}\left(\frac{2}{5x}\right)\right) + c$$

$$= \sqrt{25x^2-4} - 2\cos^{-1}\left(\frac{2}{5x}\right) + c$$

We now have the answer back in terms of x.

This first one needed lot's of explanation since it was the first one. The remaining examples won't need quite as much explanation and so won't take as long to work. However, before we move onto more problems let's first address the issue of definite integrals and how the process differs in these cases.

Example: Evaluate the following integral.

$$\int_{\frac{2}{5}}^{\frac{4}{5}} \frac{\sqrt{25x^2-4}}{x}dx$$

Solution: The limits here won't change the substitution so that will remain the same.

$$x = \frac{2}{5}\sec\theta$$

Using this substitution the square root still reduces down to,

$$\sqrt{25x^2-4} = 2|\tan\theta|$$

However, unlike the previous example we can't just drop the absolute value bars. In this case we've got limits on the integral and so we can use the limits as well as the substitution to determine the range of that we're in. Once we've got that we can determine how to drop the absolute value bars.

Here's the limits of θ.

$$x = \frac{2}{5} \quad \Rightarrow \quad \frac{2}{5} = \frac{2}{5}\sec\theta \quad \Rightarrow \quad \theta = 0$$

$$x = \frac{4}{5} \quad \Rightarrow \quad \frac{4}{5} = \frac{2}{5}\sec\theta \quad \Rightarrow \quad \theta = \frac{\pi}{3}$$

So, if we are in the range $\frac{2}{5} \leq x \leq \frac{4}{5}$ then θ is in the range of $0 \leq \theta \leq \frac{\pi}{3}$ and in this range of θs tangent is positive and so we can just drop the absolute value bars.

Let's do the substitution. Note that the work is identical to the previous example and so most of it is left out. We'll pick up at the final integral and then do the substitution.

$$\begin{aligned} \int_{\frac{2}{5}}^{\frac{4}{5}} \frac{\sqrt{25x^2-4}}{x} dx &= 2\int_0^{\frac{\pi}{3}} \sec^2\theta - 1 d\theta \\ &= 2\left(\tan\theta - \theta\right)\Big|_0^{\pi/3} \\ &= 2\sqrt{3} - \frac{2\pi}{3} \end{aligned}$$

Note that because of the limits we didn't need to resort to a right triangle to complete the problem.

Let's take a look at a different set of limits for this integral.

Example: Evaluate the following integral.

$$\int_{-\frac{4}{5}}^{-\frac{2}{5}} \frac{\sqrt{25x^2-4}}{x} dx$$

Solution: Again, the substitution and square root are the same as the first two examples.

$$x = \frac{2}{5}\sec\theta \qquad \sqrt{25x^2-4} = 2\left|\tan\theta\right|$$

Let's next see the limits for this problem.

$$x = -\frac{2}{5} \quad \Rightarrow \quad -\frac{2}{5} = \frac{2}{5}\sec\theta \quad \Rightarrow \quad \theta = \pi$$

$$x = -\frac{4}{5} \quad \Rightarrow \quad -\frac{4}{5} = \frac{2}{5}\sec\theta \quad \Rightarrow \quad \theta = \frac{2\pi}{3}$$

Note that in determining the value of θ we used the smallest positive value. Now for this range of *x*'s we have $\frac{2\pi}{3} \le \theta \le \pi$ and in this range of θ tangent is negative and so in this case we can drop the absolute value bars, but will need to add in a minus sign upon doing so. In other words,

$$\sqrt{25x^2 - 4} = -2\tan\theta$$

So, the only change this will make in the integration process is to put a minus sign in front of the integral. The integral is then,

$$\int_{\frac{2}{5}}^{\frac{4}{5}} \frac{\sqrt{25x^2 - 4}}{x} dx = -2\int_0^{\frac{\pi}{3}} \sec^2\theta - 1 d\theta$$

$$= 2\left(\tan\theta - \theta\right)\Big|_{2\pi/3}^{\pi}$$

$$= \frac{2\pi}{3} - 2\sqrt{3}$$

In the last two examples we saw that we have to be very careful with definite integrals. We need to make sure that we determine the limits on θ and whether or not this will mean that we can drop the absolute value bars or if we need to add in a minus sign when we drop them. Before moving on to the next example let's get the general form for the substitution that we used in the previous set of examples.

$$\sqrt{b^2x^2 - a^2} \qquad \Rightarrow \qquad x = \frac{a}{b}\sec\theta$$

Let's work a new and different type of example.

Example: Evaluate the following integral.

$$\int \frac{1}{x^4\sqrt{9 - x^2}} dx$$

Solution: Now, the square root in this problem looks to be (almost) the same as the previous ones so let's try the same type of substitution and see if it will work here as well.

$$x = 3\sec\theta$$

Using this substitution the square root becomes,

$$\sqrt{9 - x^2} = \sqrt{9 - 9\sec^2\theta} = 3\sqrt{1 - \sec^2\theta} = 3\sqrt{-\tan^2\theta}$$

So using this substitution we will end up with a negative quantity (the tangent squared is always positive of course) under the square root and this will be trouble. Using this substitution will give complex values and we don't want that. So, using secant for the substitution won't work.

However, the following substitution (and differential) will work.

$$x = 3 \sin \theta$$
$$dx = 3 \cos \theta d\theta$$

With this substitution the square root is,

$$\sqrt{9-x^2} = 3\sqrt{1-\sec^2\theta} = 3\sqrt{\cos^2\theta} = 3|\cos\theta| = 3\cos\theta$$

We were able to drop the absolute value bars because we are doing an indefinite integral and so we'll assume that everything is positive. The integral is now,

$$\int \frac{1}{x^4\sqrt{9-x^2}}dx = \int \frac{1}{81\sin^4\theta(3\cos\theta)}3\cos\theta d\theta$$
$$= \frac{1}{81}\int \frac{1}{\sin^4\theta}d\theta$$
$$= \frac{1}{81}\int \csc^4\theta d\theta$$

In the previous section we saw how to deal with integrals in which the exponent on the secant was even and since cosecants behave an awful lot like secants we should be able to do something similar with this.

Here is the integral.

$$\int \frac{1}{x^4\sqrt{9-x^2}}dx = \frac{1}{81}\int \csc^2\theta\csc^2\theta d\theta$$
$$= \frac{1}{81}\int (\cot^2\theta+1)\csc^2\theta d\theta \qquad u = \cot\theta$$
$$= \frac{1}{81}\int u^2+1du$$
$$= -\frac{1}{81}\left(\frac{1}{3}\cot^3\theta+\cot\theta\right)+c$$

Now we need to go back to x's using a right triangle. Here is the right triangle for this problem and trig functions for this problem.

$$\sin\theta = \frac{x}{3} \qquad \cos = \frac{\sqrt{9-x^2}}{x}\theta$$

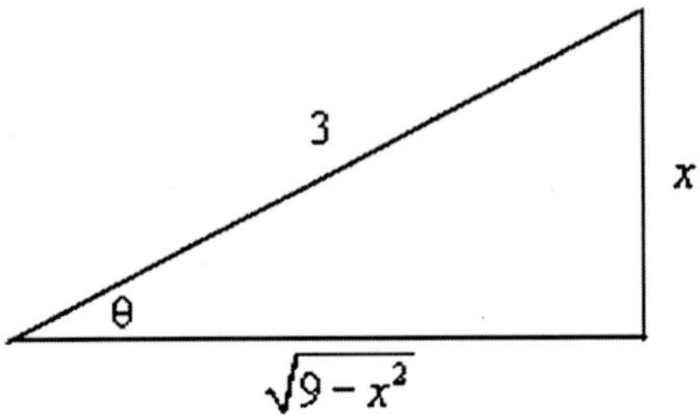

The integral is then,

$$\int \frac{1}{x^4\sqrt{9-x^2}}\,dx = -\frac{1}{81}\left(\frac{1}{3}\left(\frac{\sqrt{9-x^2}}{x}\right)^3 + \frac{\sqrt{9-x^2}}{x}\right) + c$$

$$= -\frac{(9-x^2)^{\frac{3}{2}}}{243x^3} - \frac{\sqrt{9-x^2}}{81x} + c$$

Here's the general form for this type of square root.

$$\sqrt{a^2 - b^2x^2} \qquad \Rightarrow \qquad x = \frac{a}{b}\sin\theta$$

There is one final case that we need to look at. The next integral will also contain something that we need to make sure we can deal with.

Example: Evaluate the following integral.

$$\int_0^{\frac{1}{6}} \frac{x^5}{\left(36x^2+1\right)^{\frac{3}{2}}}\,dx$$

Solution: First, notice that there really is a square root in this problem even though it isn't explicitly written out. To see the root let's rewrite things a little.

$$\left(36x^2+1\right)^{\frac{3}{2}} = \left(\left(36x^2+1\right)^{\frac{1}{2}}\right)^3 = \left(\sqrt{36x^2+1}\right)^3$$

This square root is not in the form we saw in the previous examples. Here we will use the substitution for this root.

$$x = \frac{1}{2}\tan\theta \qquad dx = \frac{1}{6}\sec^2\theta d\theta$$

With this substitution the denominator becomes,

$$\left(36x^2+1\right)^3 = \left(\sqrt{\tan^2\theta+1}\right)^3 = \left(\sqrt{\sec^2\theta}\right)^3 = \left|\sec\theta\right|^3$$

Now, because we have limits we'll need to convert them to θ so we can determine how to drop the absolute value bars.

$$x = 0 \quad \Rightarrow \quad 0 = \frac{1}{6}\tan\theta \quad \Rightarrow \quad \theta = 0$$

$$x = \frac{1}{6} \quad \Rightarrow \quad \frac{1}{6} = \frac{1}{6}\tan\theta \quad \Rightarrow \quad \theta = \frac{\pi}{4}$$

In this range of θ secant is positive and so we can drop the absolute value bars.

Here is the integral,

$$\int_0^{\frac{1}{6}} \frac{x^5}{(36x^2+1)^{\frac{3}{2}}} dx - \int_0^{\frac{\pi}{4}} \frac{\frac{1}{7776}\tan^5\theta}{\sec^2\theta}\left(\frac{1}{6}\sec^2\theta\right) d\theta$$

$$-\frac{1}{46656}\int_0^{\frac{\pi}{4}} \frac{\tan^5\theta}{\sec\theta} d\theta$$

There are several ways to proceed from this point. Normally with an odd exponent on the tangent we would strip one of them out and convert to secants.

However, that would require that we also have a secant in the numerator which we don't have. Therefore, it seems like the best way to do this one would be to convert the integrand to sines and cosines.

$$\int_0^{\frac{1}{6}} \frac{x^5}{(36x^2+1)^{\frac{3}{2}}}\,dx = \frac{1}{46656}\int_0^{\frac{\pi}{4}} \frac{\sin^5\theta}{\cos^4\theta}\,d\theta$$

$$= \frac{1}{46656}\int_0^{\frac{\pi}{4}} \frac{(1-\cos^2\theta)^2}{\cos^4\theta}\sin\theta\,d\theta$$

We can now use the substitution $u = cos\ \theta$ and we might as well convert the limits as well.

$$\theta = 0 \qquad u = \cos 0 = 1$$

$$\theta = \frac{\pi}{4} \qquad u = \cos\frac{\pi}{4} = \frac{\sqrt{2}}{2}$$

The integral is then,

$$\int_0^{\frac{1}{6}} \frac{x^5}{(36x^2+1)^{\frac{3}{2}}}\,dx = -\frac{1}{46656}\int_1^{\sqrt{2}} u^{-4} - 2u^{-2} + 1\,du$$

$$= \frac{1}{46656}\left(-\frac{1}{3u} + \frac{2}{u} + u\right)\Bigg|_1^{\frac{\sqrt{2}}{2}}$$

$$= \frac{1}{17496} - \frac{11\sqrt{2}}{279936}$$

The general form for this final type of square root is

$$\sqrt{a^2 + b^2x^2} \Rightarrow x = \frac{a}{b}\tan\theta$$

We have a couple of final examples to work in this section. Not all trig substitutions will just jump right out at us.

Sometimes we need to do a little work on the integrand first to get it into the correct form and that is the point of the remaining examples.

Chapter 14

Sequences and Series of Functions

Taylor's Theorem for an Arbitrary Function

These formulae suggest that we should seek a similar formula in the case of an arbitrary function $f(x)$, which is not necessarily a polynomial; however, in this case, the formula can lead only to an approximation to the function by a polynomial.

We wish to compare the values of the function f at the point x and at the point $\xi = x+h$, so that $h = \xi - x$. If now n is any positive integer, as a rule, the expression,

$$f(x) + (\xi - x)\, f'(x) + \ldots + \frac{(\xi - x)^n}{n!} f^{(n)}(x)$$

will not be an exact expression for the functional value $f(\xi)$, whence we must set

$$f(x) = f(x) + (x - x)f'(x) + \frac{(\xi - x)^2}{2!} f''(x)$$

$$+ \ldots \frac{(\xi - x)^n}{n!} f^{(n)}(x) + R_n,$$

where the expression R_n denotes the remainder when $f(\xi)$ is replaced by the expression $f(x) + f'(x)(\xi - x) + \ldots$ In the first instance, this equation is nothing but a formal definition of the expression R_n. Its significance lies in the fact that we can easily find a neat and useful expression for R_n. For this purpose, we think of the quantity ξ as the fixed and the quantity x as the independent variable. The remainder is then the function $R_n(x)$. By the above equation, this function vanishes for $x = \xi$:

$$R_n(x) = 0$$

Moreover, by differentiation, we obtain,

$$R_n'(x) = \frac{(\xi - x)^n}{n!} f^{(n+1)}(x)$$

In fact, if we differentiate this equation for the remainder with respect to x, we obtain 0 on the left hand side, since $f(\xi)$ does not depend on x and is therefore to be regarded as a constant. On the right hand side, we differentiate each term by the rule for products and find that all the terms cancel except the last one, which is written above with a minus sign. Now, by the fundamental theorem of the integral calculus,

$$R_n(x) = R_n(x) - R_n(x)$$

$$= \int_{\xi}^{x} R_n'(t)\,dt = -\int_{x}^{\xi} R_n'(t)\,dt,$$

so that we obtain the formula,

$$R_n(x) = \int_{x}^{x+h} \frac{(x+h-t)^n}{n!} f^{(n+1)}(t)\,dt$$

If we introduce a new integration variable τ by means of the equation $\tau = t - x$, this becomes,

$$R_n = \frac{1}{n!}\int_{0}^{h} (h-\tau)^{nf(n+1)}(x+\tau)\,d\tau$$

Collecting these results, we can make the statement:

If the function $f(x)$ has continuous derivatives up to order $(n + 1)$ in the interval under consideration, then

$$f(x+h) = f(x) + hf'(x) + \frac{h^2}{2!} f''(x) + \frac{h^3}{3!} f'''(x)$$

$$+ \dots + \frac{h^n}{n!} f^{(n)}(x) + R_n,$$

or (the equivalent expression for $h = \xi - x$)

$$f(x) = f(x) + (x - x)f'(x) + \frac{(\xi - x)^2}{2!} f''(x)$$

$$+\ldots \frac{(\xi - x)^n}{n!} f^{(n)}(x) + R_n,$$

where the remainder R_n is given by

$$R_n = \frac{1}{n!}\int_0^h (h-\tau)^n f^{(n+1)}(x+\tau)\,d\tau$$

In particular, if we set $x = 0$ and then replace h by x, we obtain

$$f(x) = f(0) + \frac{x}{1!}f'(0) + \frac{x^2}{2!}f''(0) + \ldots + \frac{x^n}{n!}f^{(n)}(0) + R_n$$

with the remainder

$$R_n = \frac{1}{n!}\int_0^x (x-\tau)^n f^{(n+1)}(\tau)\,d\tau$$

These formulae are known as Taylor's theorem. They give expressions for the functions $f(x + h)$ and $f(x)$, respectively, in terms of polynomials of degree n in h and in x, respectively (the so-called polynomial of approximation), and a remainder. The polynomial of approximation is characterized by the fact that, when $h = 0$ (or $x = 0$, as the case may be), its value and that of its first n derivatives are the same as those of the given function and its first n derivatives. In contrast to the Taylor series for polynomials, which do not require a remainder, the remainder and the expression for it are here essential.

The significance of the formula lies in the fact that the remainder, even though it has a more complicated form than the other terms of the formula, nevertheless provides useful means for estimating the accuracy with which the sum of the first $n + 1$ terms,

$$f(0) + \frac{x}{1!}f'(0) + \frac{x^2}{2!}f''(0) + \ldots + \frac{x^n}{n!}f^{(n)}(0),$$

represents the function $f(x)$.

Estimation of the Remainder: Whether the first $n + 1$ terms of Taylor's series actually yield a sufficiently good approximation to the function depends naturally on whether the remainder is sufficiently small. Hence we turn now to the estimation of this remainder. Such an estimate can most easily be made by means of the mean value theorem of the integral calculus.

We employ this theorem in the form

$$\int_0^h p(\tau)\phi(\tau)d\tau = \phi(\theta h)\int_0^h p(\tau)d\tau,$$

where $p(\tau)$ is a continuous function which is negative nowhere in the interval of integration and $\phi(\tau)$ is merely a continuous function there, while q is a number in the interval $0 < q < 1$. In fact, we may assume that $0 < \theta < 1$, but this is not important here. If we set in the expression for the remainder $p(\tau)=(h-\tau)$, we obtain

$$R_n = \frac{h^{n+1}}{(n+1)!} f^{(n+1)}(x+\theta h),$$

while if we set $p(\tau) = 1$, we obtain

$$R_n = \frac{h^{n+1}}{n!}(1-\theta)^n f^{(n+1)}(x+\theta h),$$

which is less important for us and is stated here only for the sake of completeness. In these formulae, θ denotes a certain number in the interval $0 \} \theta \} 1$, the value of which, in general, cannot be specified more accurately; of course, as a rule, this value is different in the two formulae for the remainder and depends, in addition, on n, x and h.

The first form of the remainder was given by Lagrange, the second by Cauchy, whence they received their names.

These expressions for the remainder, as well as others, can be derived from the mean value theorem of the differential calculus and from the generalized mean value theorem, respectively.

We apply these theorems to the function $R_n = R_n(x) - R_n(\xi)$ and to the pair of functions $R_n(x)$ and $(x - \xi)^{n+1}$, respectively, where we consider ξ to be fixed and employ the formula

$$R_n'(x) = -\frac{(\xi - x)^n}{n!} f^{(n+1)}(x)$$

These methods of deriving the formulae for the remainder emphasize more the fact that Taylor's theorem is a generalization of the mean value theorem; they also offer the advantage, which for many theoretical purposes is important, that we need only assume the existence and not the continuity of the $(n + 1)$-th derivative.

However, on the other hand, we lose the advantage of having an exact expression for the remainder in the form of an integral. Our interest will be directed chiefly towards discovering whether the remainder R_n tends to zero as n increases; if this is the case, the larger we choose n the more accurately is $f(x + h)$ represented by the corresponding polynomial in h. In this case, we say that we have expanded the function in an infinite Taylor series

$$f(x+h) = f(x)+\frac{h}{1!}f'(x)+\frac{h^2}{2!}f''(x)+\frac{h^3}{3!}f'''(x)+...,$$

or, in particular, if we first set $x = 0$ and then write x in place of h,

$$f(x+h) = f(0)+\frac{x}{1!}f'(0)+\frac{x^2}{2!}f''(0)+\frac{x^3}{3!}f'''(0)+....$$

We shall encounter examples in the next section. However, first of all, we wish to point out the second important point of view arising from the consideration of Taylor series.

If we think in the first formula of the quantity h as becoming smaller and tending to zero, then, in the terminology of, the various terms of the series will tend to zero at different orders of magnitude; we accordingly call the expression $f(x)$ the term of zero order in the Taylor series, the expression $hf'(x)$ the term of first order, the expression $hf''(x)/2!$ the term of second order, etc. We gather from the form of our remainder the fact: In expanding a function as far as the term of n-th order, we make an error which tends to zero at order $(n+1)$ as $h < 0$.

Many important applications depend on this fact. It shows us that the nearer the point $x + h$ lies to the point x, the better is the representation of the function $f(x+h)$ by the polynomial approximation and that, in a given case, the approximation can be improved in the immediate neighbourhood of the point x by increasing the value of n.

Expansions of the Elementary Functions

We shall now use the general results of the preceding section in order to express the elementary functions approximately by polynomials and to expand them in Taylor series. However, we shall retrict ourselves to those functions for which the coefficients of the expansion in series are given by simple laws of formation. The Exponential function:

The simplest example is offered by the exponential function $f(x) = e^x$, all the derivatives are which identical with $f(x)$ and therefore have the value 1 for $x = 0$. Hence, using Lagrange's form for the remainder.

$$e^{\alpha} = 1 + \frac{x}{1!} + \frac{x^2}{2!} + \frac{x^3}{3!} + \ldots + \frac{x^n}{n!} + \frac{x^{n+1}}{(n+1)!} e^{\theta x}$$

If we now let n increase beyond all bounds, the remainder will tend to zero, no matter what fixed value of x has been chosen, because $|e^{\theta x}| \succ e^{|x|}$ from the start.

We now choose a fixed integer m large than $2|x|$, then for $n \succ m$,

$$\frac{|x|}{n} < \frac{1}{2}; \left|\frac{x^{n+1}}{(n+1)!}\right| = \frac{|x^m|}{m!} \cdot \frac{|x|}{m+1} \cdots \frac{|x|}{n+1} \leqq \frac{|x^m|}{m!} \frac{1}{2^{n+1-m}} \leqq \frac{|2x|^m}{m!} \frac{1}{2^n},$$

$$\text{whence } |R_n| \leqq \frac{|2x|^m}{m!} e^{|x|} \frac{1}{2^n}$$

Since the first two factors on the right hand side are independent of x, while the number $1/2^n$ tends to zero as n increases, our statement is proved. If we think of the number x as not being fixed, but free to vary in the interval $-a < x < a$, where a is a fixed positive number, there follows from the above,

if we choose $m > 2a$, that the estimate, $|R_n| \leqq \frac{|2a|^m}{m!} e^a \frac{1}{2^n}$ is valid

provided $n < m$. Thus, we have specified for the remainder a bound which holds for all values of x in the interval $-a < x < a$ and tends to zero as n. For the function e^x, we can therefore write the expansion as an infinite series

$$e^x = 1+\frac{x}{1!}+\frac{x^2}{2!}+\frac{x^3}{3!}+\ldots=\sum_{\nu=0}^{\infty}\frac{x^\nu}{\nu!}$$

This expansion is valid for all values of x. Thus, we have again proved that the number e considered is the same as the base of the natural logarithm. Of course, we must employ for numerical calculations the finite form of Taylor's theorem with the remainder; for example, for $x = 1$, this yields

$$e = 1 + 1 + \frac{1}{2!}+\frac{1}{3!}+\ldots+\frac{1}{n!}+\frac{e^\theta}{(n+1)!}$$

If we wish to compute e with an error of at most 1/10,000, we need only choose n so large that the remainder is certainly less than 1/10,000; since this remainder is certainly less* than $3/(n + 1)!$, it is sufficient to choose n = 7, since 8!>30,000. We thus obtain the approximate value $e = 2.71822$ with an error less than 0.0001. We do not take here into account the error due to neglect of the figures in the sixth decimal place. We have used here the fact that $e < 3$. This follows immediately from our series for e, because it is always true that $1/n < 1/2^{n-1}$, whence

$$e < 1 + 1 + \frac{1}{2} + \frac{1}{4} + \ldots = 1 + 1/(1 - \frac{1}{2}) = 3$$

Sin x, cos x, sinh x, cosh x: We find for the functions sin x, cos x,

sinh x and cosh x:

$$\begin{array}{lcllll} f(x) & = & \sin x & \cos x & \sinh x & \cosh x, \\ f'(x) & = & \cos x & -\sin x & \cosh x & \sinh x, \\ f''(x) & = & -\sin x & -\cos x & \sinh x & \cosh x, \\ f'''(x) & = & -\cos x & \sin x & \cosh x & \sinh x, \\ f^{iv}(x) & = & \sin x & \cos x & \sinh x & \cosh x \end{array}$$

If $f(x) = \sin x$ or $f(x) = \cos x$, the n-th derivative can always be represented by,

$$f^{(\pi)}(x) = f(x + \frac{1}{2}n\pi)$$

Hence, the coefficients of the even powers of x vanish in the polynomial approximations for sin x and sinhx, while the coefficients of the odd powers vanish in those for cos x and cosh x. Thus, in the first case, the $(2n + 1)$-th and the (2n + 2)-th polynomial are identical, while, in the second case, the $2n$-th and the $(2n + 1)$-th polynomials are identical. If we employ in each case the higher order ones of these polynomials, we obtain at once Lagrange's forms of the remainders

$$\sin x = x - \frac{x^3}{3!} + \frac{x^5}{5!} - + \ldots + (-1)^n \frac{x^{2n+1}}{(2n+1)!} + (-1)^{n+1} \frac{x^{2n+3}}{(2n+3)!} \cos(\theta x)$$

$$\cos x = 1 - \frac{x^2}{2!} + \frac{x^4}{4!} - + \ldots + (-1)^n \frac{x^{2n}}{(2n)!} + (-1)^{n+1} \frac{x^{2n+2}}{(2n+2)!} \cos(\theta x),$$

$$\sinh x = x + \frac{x^3}{3!} + \frac{x^5}{5!} - + \ldots + \frac{x^{2n+1}}{(2n+1)!} + \frac{x^{2n+3}}{(2n+3)!} \cosh(\theta x),$$

$$\cosh x = 1 + \frac{x^2}{2!} + \frac{x^4}{4!} + \ldots + \frac{x^{2n}}{(2n)!} + \frac{x^{2n+2}}{(2n+2)!} \cosh(\theta x),$$

where, of course, in each of the four formulae θ denotes a different number in the interval $0 < \theta < 1$, a number which, in addition, depends on n and on x.

In these formulae, we can also make the approximation as exact as we please for each value of x, since the remainder tends to 0 as n increases. We thus obtain the four series

$$\sin x = x - \frac{x^3}{3!} + \frac{x^5}{5!} - + \ldots = \sum_{\nu=0}^{\infty} (-1)^\nu \frac{x^{2\nu} + 1}{(2\nu+1)!},$$

$$\cos x = 1 - \frac{x^2}{2!} + \frac{x^4}{4!} - + \ldots = \sum_{\nu=0}^{\infty} (-1)^\nu \frac{x^{2\nu}}{(2\nu)!},$$

$$\sinh x = x + \frac{x^3}{3!} + \frac{x^5}{5!} + \ldots = \sum_{\nu=0}^{\infty} \frac{x^{2\nu+1}}{(2\nu+1)!},$$

$$\cosh x = 1 + \frac{x^2}{2!} + \frac{x^4}{4!} + \ldots = \sum_{\nu=0}^{\infty} \frac{x^{2\nu}}{(2\nu)!}$$

The last two series can also be obtained formally from the series for e^x in accordance with the definitions of the hyperbolic functions.

The Binomial Series

We may pass over the Taylor series for the functions log $(1 + x)$ and artan x, which we have already dealt with directly in equation. However, we must take up the generalization of the binomial theorem for arbitrary indices, which is one of the most fruitful of Newton's mathematical discoveries and which represents one of the most important cases of the expansion in Taylor series. Our objective is the expansion of the function

$$f(x) = (1 + x)^\alpha$$

in a Taylor series, where $x > -1$ and α is an arbitrary number, positive or negative, rational or irrational. We have chosen the function $(1 + x)^\alpha$ instead of x^α, because it is not true at the point $x = 0$ that all the derivatives of x^α are continuous except in the trivial case of non-negative values of α.

We calculate first the derivatives of f(x):

$$f'(x) = \alpha(1 + x)^{\alpha-1},\ f''(x) = \alpha(\alpha - 1)(1 + x)^{\alpha-2},\ldots,$$

$$f^{(\nu)}(x) = \alpha(\alpha - 1)\ldots(\alpha - \nu + 1)(1 + x)^{\alpha-\nu}$$

In particular, we have for $x = 0$,

$$f'(0) = \alpha,\ f''(0) = \alpha(\alpha - 1),\ldots,\ f^{(\nu)}(0)$$

$$= \alpha(\alpha - 1)\ \alpha(\alpha - \nu + 1)$$

Taylor's theorem then yields

$$(1 + x)^\alpha = 1 + ax + \frac{\alpha(\alpha - 1)}{2!}x^2$$

$$+ \ldots + \frac{\alpha(\alpha - 1)(\alpha - 2)\ldots(\alpha - n + 1)}{n!}x^n + R_n$$

We must still discuss the remainder. This is not very difficult, but nevertheless is not quite as simple as in the cases treated previously. The result, which we mention here in advance, is that whenever $|x| < 1$, the remainder tends to 0, whence the expression $(1 + x)^\alpha$ can be expanded in the infinite binomial series

$$\left(1+x\right)^{\alpha} = 1 + \frac{\alpha}{1!}x + \frac{\alpha\left(\alpha-1\right)}{2!}x^2 + \ldots = \sum_{v=0}^{\infty}\binom{\alpha}{v}x^y$$

where, for the sake brevity, we have introduced the general binomial coefficients

$$\binom{\alpha}{v} = \frac{\alpha\left(\alpha-1\right)\ldots\left(\alpha-v+1\right)}{v!}\left(for\,v>0\right),\binom{\alpha}{0}=1$$

Geometrical Applications

The behaviour of a function $f(x)$ in the neighbourhood of a point $x = a$ or of a given curve in the neighbourhood of a point can be studied with increased accuracy by means of Tailor's theorem, because this theorem resolves the increment of the function on passing to a neighbouring point $x = a + h$ into a sum of quantities of the first order, second order, etc.

Contact of Curves

We shall now use this method, in order to investigate the concept of contact of two curves. If at a point, say the point $x = a$, two curves $y = f(x)$ and $y = g(x)$ do not only intersect, but also have a common tangent, we shall say that at this point the curves touch one another or have contact of the first order.

The Taylor expansions of the functions $f(a + h)$ and $g(a + h)$ then have the same terms of zero and first order in h.

If at the point $x=a$ the second derivatives of $f(x)$ and $g(x)$ are also equal to each other, we say that the curves have contact of the second order.

In the Taylor expansions, the terms of second order are then also the same and, if we assume that both functions have at least continuous derivatives of the third order, the difference $D(x) = f(x) - g(x)$ can be expressed in the form

$$D(a + h) = \quad f(a + h) - g(a + h) = \frac{h^3}{3!}\, D'''(a + qh)$$

where the expression $F(h)$ tends to $f'''(a + h)$ - g'''(a) as h tends to zero, whence the difference $D(a + h)$ vanishes to at least the third order in h.

We can proceed in this way and consider the general case, when the Taylor series for $f(x)$ and $g(x)$ are the same up to terms of the n-th order, i.e., when

$$f(a) = g(a),\ f'(a) = g'(a),\ f''(a)= g''(a),\dots,f^{(n)}(a) = g^{(n)}(a)$$

We assume here that the $(n + 1)$-th derivatives are also continuous. Under these conditions, we say that at this point the curves have contact of the n-th order. The difference of the two functions will then be

$$f(a + h) - g(a + h) = \frac{h^{n+1}}{(n+1)} f(h)$$

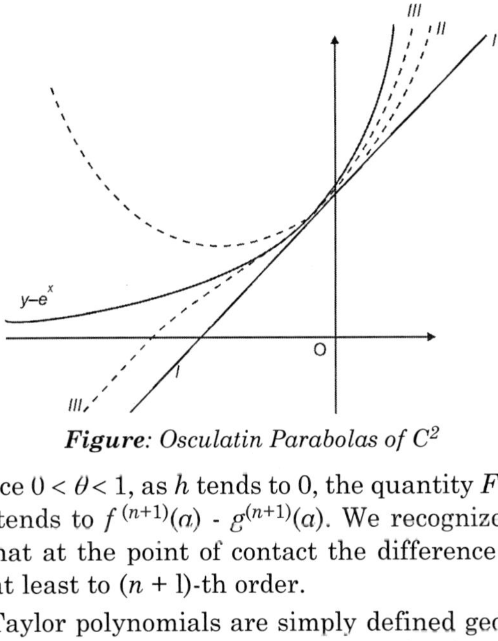

Figure: *Osculatin Parabolas of* C^2

where, since $0 < \theta < 1$, as h tends to 0, the quantity $F(h) = D^{(n+1)}(a + q\ h)$ tends to $f^{(n+1)}(a) - g^{(n+1)}(a)$. We recognize from this formula that at the point of contact the difference $f(g) - g(x)$ vanishes at least to $(n + 1)$-th order.

The Taylor polynomials are simply defined geometrically by the fact that they are those parabolas of order n which at the given point have contact of the highest possible order with the graph of the given function. Hence they are sometimes called osculating parabolas. For the case $y = e^x$, figures gives the first three osculating parabolas at the point $x = 0$.

If two curves $y = f(x)$ and $y = g(x)$ have contact of order n, the definition does not exclude the possibility that the contact may be of a still higher order, i.e., that also the equation $f^{(n+1)}(a) = g^{(n+1)}(a)$ is true.

If this is not the case, i.e., if $f^{(n+1)}(a)$ ýÿ $g^{(n+1)}(a)$, we speak of a contact of exactly the n-th order or say that the order of the contact is exactly n.

The fact that the order of contact of two curves is a genuine geometrical relationship, which is unaffected by a change of axes, is easily verified by means of the formulae for a change of axes.

From our formula as well as from our figures, we can at once state a remarkable fact which is often overlooked by beginners.

If the contact of two curves is exactly of an even order, that is, if an even number n of derivatives of the two functions have the same value at the point in question, while the $(n+1)$-th derivatives differ, then, in conformity with the above formula, the difference $f(a + h) - g(a + h)$ will have different signs for small positive values and for numerically small negative values of h.

The two curves will then cross at the point of contact. This case occurs, for example, in a contact of the second order, if the third derivatives have different values.

However, if we consider in detail the case of an odd order contact, e.g., the case of an ordinary contact of the first order, the difference $f(a + h) - g(a + h)$ will have the same sign for all numerically small values of h, whether positive or negative; the two curves therefore will not cross in the neighbourhood of the point of contact.

The simplest example of this type is the contact of a curve with its tangent.

The tangent can cross the curve only at points where the contact is at least of second order; it will actually cross the curve at points, where the order of contact is even, e.g., at an ordinary point of inflection, where f “(x) = 0, but f”(x) ýÿ 0.

At points, where the order of contact is odd, it will not cross the curve; examples are an ordinary point of the curve where the second derivative is non-zero or the curve $y = x^4$ at the origin.

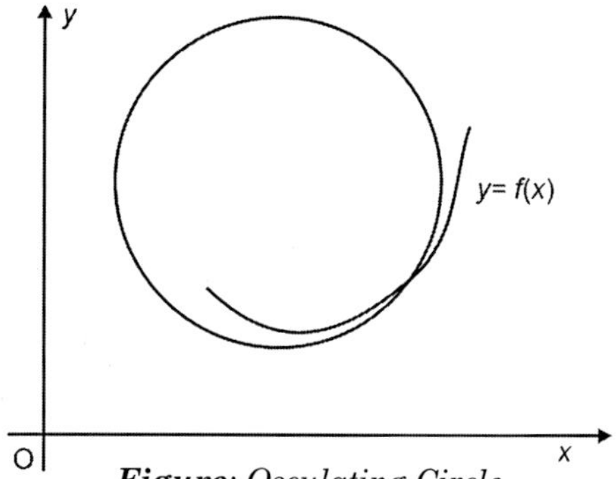

***Figure:** Osculating Circle*

The Circle of Curvature as Osculating Circle:. From this point of view, the concept of the curvature of a curve $y = f(x)$ gains a new intuitive significance. There pass through the definite point of the curve with co-ordinates $x = a$ and $y = b$ an infinite number of circles which touch the curve at the point. The centres of these circles lie on the normal to the curve and there corresponds to each point of this normal just one such tangent circle. We may expect that we can bring about by a proper choice of the centre of the circle a contact of second order between the curve and the circle.

As a matter of fact, we know from that for the circle of curvature at the point $x = a$, the equation of which is, say, $y = g(x)$, we do not only have $g(a) = f(a)$ and $g'(a) = f'(a)$, but also $g''(a) = f''(a)$, whence the circle of curvature is at the same time the osculating circle at the point of the curve under consideration, i.e., it is the circle which at that point has second order contact with the curve. In the limiting case of a point of inflection or, in general, of a point at which the curvature is zero and the radius of curvature infinite, the circle of curvature degenerates into the tangent. In ordinary cases, i.e., when the contact at the point in question does not happen to be of an order higher than the second, the circle of curvature will not merely touch the curve, but will also cross it.

On the Theory of Maxima and Minima

A point $x = a$, at which $f'(a) = 0$, has a maximum of the function $f(x)$ if $f''(a)$ is negative, a minimum if $f''(a)$ is positive,whence these conditions are sufficient for the occurrence of a maximum or a minimum. They are by no means necessary, because there are three possibilities in the case where $f''(a) = 0$; at the point in question, the function may have a maximum or a minimum or neither. Examples of the three possibilities are giver by the functions $y=-x^4$, $y=x^4$ and $y = x^3$ at the point $x = 0$. Taylor's theorem enables us immediately to make a general statement of sufficient conditions for a maximum or a minimum.We merely need expand the function $f(a + h)$ in powers of h; the essential point is then to discover whether the first non-vanishing term contains an even or an odd power of h. In the first case, we have a maximum or a minimum according to whether the coefficient of h is negative or positive; in the second case, we have a horizontal inflectional tangent and neither a maximum nor a minimum. The reader should complete the argument by using the formula for the remainder. However, the earlier necessary and sufficient condition is more general and more convenient in applications, i.e., provided the first derivative $f'(x)$ vanishes at only a finite number of points; a necessary and sufficient condition for the occurrence of a maximum or minimu,at one of these points is that the first derivative $f'(x)$ changes sign as the point is passed.

infinite series

The terms of the infinite series hitherto considered have been constants, whence these series (when convergent) always represented definite numbers. But both in theory and in applications, the series of outstanding importance are those in which the terms are functions of a variable, so that the sum of the series is also a function of the variable, as in the case of Taylor series. We shall therefore consider a series $g_1(x) + g_2(x) + g_s(x) + \ldots$, in which the functions $g_n(x)$ are defined in an interval $a > x > b$. We will denote the n-th partial sum $g_1(x) + g_2(x) + \ldots + g_n(x)$, by $f_n(x)$. Then the sum $f(x)$ of our series, where it exists, is simply the limit $\lim_{n\to\infty} f_n(x)$. We may therefore regard the sum of an infinite series of functions as the limit of a sequence of functions $f_1(x)$, $f_2(x)$, $f_n(x)$,... Conversely, for any such sequence of functions $f_1(x)$, $f_2(x)$,..., we can form an equivalent series by setting $g_1(x) = f_1(x)\ldots$ and $g_n(x) = f_n(x) - f_{n-1}(x)$ for $n > 1$. Hence, when it is convenient, we can pass from the consideration of series to that of sequences and conversely.

Chapter 15

Elementary Integrals

The Method of Substitution

The first of these useful methods for attacking integration problems is the introduction of a new variable (i.e. the method of substitution or transformation). The corresponding integral formula is just the chain rule of the differential calculus, expressed in the integral form.

The Substitution Formula

We assume that a new variable u is introduced into a function $F(x)$ by means of the equation $x = \phi(u)$, so that $F(x)$ becomes a function of u:

$$F(x) = F(\phi(u)) = G(u)$$

By the chain rule of the differential calculus,

$$\frac{dG}{du} = \frac{dF}{dx}\phi'(u)$$

If we now write

$$F'(x) = f(x)$$

and

$$G'(u) = g(u)$$

or the equivalent expressions

$$F(x) = \int f(x)dx$$

and

$$G(u) = \int g(u)du$$

then, on the one hand, the chain rule takes the form

$$g(u) = \int(x)\phi'(u)$$

and, on the other hand, $G(u) = F(x)$ by definition, that is,

$$\int g(u)\,du = \int f(x)\,dx$$

and we obtain the integral formula, equivalent to the chain rule,

$$\int f(\phi(u))\phi'(u)\,du = \int f(x)\,dx, \{x = \phi(u)\}$$

This is the basic formula for the substitution of a new variable in an integral. It means that, if we wish to find an indefinite integral of a function of u, which is given in the special form $f(\phi(u))\phi'(u)$, we can instead find the indefinite integral of the function $f(x)$ as a function of x and after integration return to the variable u by setting $x = \phi(u)$. For example, if we apply the formula to the integrand $\phi'(u)/\phi(u)$, we obtain

$$\int \frac{\phi'(u)}{\phi(u)}\,du = \int \frac{dx}{x} = \log|x| = \log|\phi(u)|$$

or, replacing u by x,

$$\int \frac{\phi'(x)}{\phi(x)}\,dx = \log|\phi(x)|$$

If we substitute in this important formula particular functions such as $\phi(x) = \log x$ or $\phi(x) = \sin x$, we obtain

$$\int \frac{dx}{x \log x} = \log|\log x|$$

$$\int \cot x\,dx = \log|\sin x|, \int \tan x\,dx = -\log|\cos x|$$

A further example is

$$\int \varphi(u)\varphi'(u)\,du = \int x\,dx = \frac{1}{2}x^2 = \frac{1}{2}[\varphi(u)]^2$$

where $f(x) = x$. This yields for $\phi(u) = \log u$

$$\int \frac{\log u}{u} du = \frac{1}{2}(\log u)^2$$

These and the following formulae are verified by showing that differentiation of the result returns the integrand. Moreover, the formulae are, of course, only asserted as true in as far as the expressions occurring in them have a meaning. We finally consider the example

$$\int \sin^n u \cos u du$$

Here $x = \sin u = \phi(u)$, whence

$$\int \sin^n u \cos u du = \int x^n dx = \frac{x^{n+1}}{n+1} = \frac{\sin^{n+1} u}{n+1}$$

However, in many cases, we shall use the above formula in the reverse direction, starting with the right hand side - the integral $\int f(x) dx$.

We now have to evaluate or simplify a prescribed indefinite integral $F(x) = \int f(x) dx$ by introduction of the new integration variable u by means of the transformation formula $x = \phi(u)$, then working out the indefinite integral

$$G(u) = \int f(\phi(u))\phi'(u) du$$

and finally replacing the variable u in this integral by x. In order to carry out this last step, we must be certain that a definite value u actually corresponds to the value x, i.e., that the function $x = \phi(u)$ has an inverse. Accordingly, we now make the following assumption, in which we regard x as the primary variable:

In the interval under consideration, $u = \psi(x)$ is a monotonic, differentiable function, the derivative $f'(x)$ of which does not vanish anywhere in the interval.

We denote the inverse function, which under these conditions is definite and single-valued, by $x = \phi(u)$; its derivative is then given by $\phi'(u) == 1/\psi'(x)$. As the basic formula for the substitution of a new variable u in an integral, we obtain

$$\int f(x)dx = \int f(\phi(u))\phi'(u)du \quad (u = \psi(x))$$

The indefinite integral $\int f(x)dx$ can be obtained by calculation of the indefinite integral $\int f(\phi(u))\phi'(u)du$ and finally introducing x instead of u for the independent variable by means of the equation $u = \psi(x)$.

Hence it is not sufficient merely to express the old variable x in terms of the new variable u and then to integrate with respect to this new variable; before integrating, we must multiply by this derivative of the original variable x with respect to the new variable u. The corresponding formula for definite integration between two limits is

$$\int_a^b f(x)dx = \int_{\psi(a)}^{\psi(b)} f(\phi(u))\phi'(u)du$$

In the new integral, we have to choose those limits of integration which are obtained by subjecting the old integration limits to the transformation $x = \phi(u)$, $u = \psi(x)$. In most applications, the integrand $f(x)$ will appear at the outset as a function of a function, say, $f(x) = h(u)$, where $u = \psi(x)$. It is then more convenient to write our integral formula in a slightly different form by identifying the expression $f\{\phi(u)\}$ with the expression $h(h)$. If we make for u the substitution $u = \psi(x)$, $x = \phi(u)$, then our transformation formula is simply

$$\int h\{\psi(x)\}dx = \int h(u)\frac{dx}{du}du$$

As a first example, consider the integration of the function $f(x) = \sin 2x$ and introduce $u = \psi(x) = 2x$ and $h(u) = \sin u$. We have

$$\frac{du}{dx} = \psi'(x) = 2$$

If we now substitute $u = 2x$ into the integral as the new variable, then it is not transformed into $\int \sin u du$, but into

$$\frac{1}{2}\int \sin u du = -\frac{1}{2}\cos u = -\frac{1}{2}\cos 2\pi$$

of course, this may of be verified at once by differentiating the right hand side.

If we integrate for x between the limits 0 and $\pi/4$, the corresponding limits for u are 0 and $\pi/2$, and we obtain

$$\int_0^{\pi/4} \sin 2x dx = \frac{1}{2}\int_0^{\pi/2} \sin u du = -\frac{1}{2}\cos u\Big|_0^{\pi/2} = \frac{1}{2}$$

Another simple example is the integral $\int_1^4 \frac{dx}{\sqrt{x}}$. Here we take

$u = \psi(x) = \sqrt{x}$ whence $x = \phi(u) = u$. Since $\phi'(u) = 2u$, we find

$$\int_1^4 \frac{dx}{\sqrt{x}} = 2\int_1^2 \frac{udu}{u} = 2\int_1^2 du = 2$$

Proof of the Substitution Formula

Our integration formula can also be explained in another and more direct manner by aiming at the formula for definite integration and basing the proof on the meaning of the definite integral as a limit of a sum. In order to calculate the integral

$$\int_a^b h(\psi(x))dx$$

(for the case $a < b$), we begin with an arbitrary subdivision of the interval $a < x < b$, and then make the sub-division finer and finer. We choose these sub-divisions in the following way.

If the function $u = \psi(x)$ is assumed to be monotonic increasing, there is a (1,1) correspondence between the interval $a < x < b$ on the x-axis and an interval $\alpha < u < \beta$ of the values of $u = \psi(z)$, where $\alpha = \psi(a)$, $\beta = \psi(b)$. We subdivide this interval into n parts of length Δu; there is a corresponding subdivision of the x-interval into subintervals which, in general, are not all of the same length. (The assumption that these subintervals are all equal it by no means essential for the proof!) We denote the points of division of the x-interval by $x_0 = a, x_1, x_2, \ldots, x_n = b$ and the lengths of the corresponding sub-intervals by

$$\Delta x_1, \Delta x_2, \ldots, \Delta x_n$$

The integral under consideration is then the limit of the sum

$$\sum_{\nu=1}^{n} h\{\psi(\xi_\nu)\}\Delta x_\nu$$

where the value ξ_ν is arbitrarily selected from the ν-th subinterval of the x-subdivision. This limit exists (for $\Delta u \to 0$) and is the integral, since on account of the uniform continuity of $x = \phi(u)$ the greatest of the lengths Δx tends to 0 with Δu.

We now write this sum in the form,

$$\sum_{\nu=1}^{n} h(u_\nu)\frac{\Delta x_\nu}{\Delta u}\Delta u \text{ where } u_\nu = \psi(\xi_\nu).$$

By the mean value theorem of the differential calculus,

$$\frac{\Delta x_\nu}{\Delta u} = \phi'(\eta_\nu)$$

where η_ν is a suitably chosen intermediate value of the variable u in the ν-th sub-interval of the u-sub-division and $x = \phi(u)$ denotes the inverse function of $u = \psi(x)$. If we now select the value ξ_ν in such a way that ξ and η coincide, i.e., $\xi_\nu = \phi(\eta_\nu)$, $\eta_\nu = \psi(\xi_\nu)$, then our sum takes the form

$$\sum_{\nu=1}^{n} h(\eta_\nu)\phi'(\eta_\nu)\Delta u$$

If we make here the passage to the limit, we immediately obtain the expression

$$\int_\alpha^\beta h(u)\frac{dx}{du}du$$

as the limiting value, that is, as the value of the integral under consideration in agreement with the formula given above.

Hence we have proved the theorem: Let $h(u)$ be a continuous function of u in the interval $\alpha < u < \beta$. If the function $u = \psi(x)$ is continuous and monotonic, has a continuous, non-vanishing derivative du/dx in $a < x < b$, and $\psi(a) = \alpha, \psi(b) = \beta$, then

$$\int_a^b h(\psi(x))dx = \int_a^b h(u)dx = \int_\alpha^\beta h(u)\frac{dx}{du}du$$

This formula exhibits the advantage of Leibnitz's notation. In order to carry out the substitution $u=\psi(x)$, we only need write $(dx/du)du$ in place of dx, changing the limits from the original values of x to the corresponding values of u.

Integration Formulae

With the help of the substitution rule, we canevaluate a given integral $\int f(x)dx$ in many cases by reducing it by means of a suitable substitution $x = \phi(u)$ to one of the elementary integrations in our table.

Whether such substitutions exist and how to find them are questions which have no general answer; this is rather a matter in which practice and ingenuity, in contrast to a systematic method, come into their own.

As an example, we shall work out the integral,

$$\int \frac{dx}{\sqrt{\left(a^2 - x^2\right)}}$$

using the substitution $x = \phi(u) = au$, $u=\psi(x)=x/a$, $dx=adu$, by which, we obtain,

$$\int \frac{dx}{\sqrt{\left(a^2 - x^2\right)}} = \int \frac{adu}{a\sqrt{\left(1-u^2\right)}} = \text{arc sinu} = \text{arc}\sin\frac{x}{a}, \text{ for}$$

$|x| < |a|$

For the sake of brevity, we have written the symbols dx and du separately, i.e., $dx = \phi'(u)du$ instead of $dx/du = \phi'(u)$.

The same substitution yields also

$$\int \frac{dx}{a^2 + x^2} = \int \frac{adu}{a^2\left(1+u^2\right)} = \frac{1}{a}\arctan u = \frac{1}{a}\arctan\frac{x}{a}$$

$$\int \frac{dx}{\sqrt{\left(a^2 + x^2\right)}} = ar\sinh\frac{x}{a}$$

$$\int \frac{dx}{\sqrt{\left(x^2 - a^2\right)}} = ar\cosh\frac{x}{a}, \text{for } |x| > |a|$$

$$\int \frac{dx}{a^2 - x^2} = \begin{cases} \frac{1}{a} ar\tanh\frac{x}{a} \text{for} |x| < |a| \\ \frac{1}{a} ar\coth\frac{x}{a} \text{for} |x| > |a| \end{cases}$$

formulae which occur very frequently and which are easily verified by differentiating the right hand side.

In conclusion, we again emphasize the point: In our substitution process, we have made the assumption that the substitution has a unique inverse $x = \phi(u)$, and indeed that $\psi'(x)$ is nowhere equal to zero in the interval under consideration. If our assumption is not fulfilled, application of the substitution formula may easily lead to wrong conclusions.

If $\psi'(x) = 0$ at isolated points of the interval of integration only, we can avoid these difficulties by subdividing this interval in such a way that $\psi'(x)$ vanishes only at the ends of a sub-interval; we can then apply the substitution to each sub-interval separately. An application of this method leads at once to the following result, which applies to many special cases: If the derivative $\psi'(x)$ vanishes at a finite number of points, but the function $\psi(x)$remains monotonic, then the substitution formula remains valid.

Further Examples of the Substitution Method

We collect here several examples which the reader should study carefully by way of practice!

By the substitution $u = 1 \} x$, $du = \} 2xdx$, we find

$$\int \frac{xdx}{\sqrt{\left(1 \pm x^2\right)}} = \pm\sqrt{\left(1 \pm x^2\right)}$$

$$\int \frac{xdx}{1 \pm x^2} = \pm\frac{1}{2}\log\left|1 \pm x^2\right|$$

In these formulae, we must take either the sign + or the sign - in all three places. By the substitution $u = ax + b$, $du = adx$ ($a \neq 0$), we find,

$$\int \frac{dx}{ax+b} = \frac{1}{a}\log|ax+b|$$

$$\int (ax+b)^{\alpha}\,dx = \frac{1}{a(\alpha+1)}(ax+b)^{\alpha+1} \qquad (\alpha \neq -1)$$

$$\int \sin(ax+b)\,dx = -\frac{1}{a}\cos(ax+b)$$

in a similar manner, using the substitution $u = \cos x$, $du = -\sin x\, dx$, we obtain

$$\int \tan x dx = -\log|\cos x|$$

and by means of the substitution $u = \sin x$, $du = \cos x\, dx$

$$\int \cot x dx = \log|\sin x|$$

Using the analogous substitutions $u = \cosh x$, $du = \sinh x\, dx$ and $u = \sinh x$, $du = \cosh x dx$, we obtain

$$\int \tanh x dx = \log|\cosh x|$$

$$\log|\sinh x| = \int \coth x dx$$

By the substitution $u = (a/b)\tan x$, $du = (a/b)\sec x\, dx$, we find the two formulae

$$\frac{1}{b^2}\int \frac{1}{a^2/b^2\tan^2 x+1}\cdot\frac{dx}{\cos^2 x} = \frac{1}{ab}\arctan\left(\frac{u}{b}\tan x\right)$$

and
$$S_m(\alpha) = \frac{1}{m+1}\left[\frac{\sin\frac{(m+1)\alpha}{2}}{\sin\frac{\alpha}{2}}\right]^2$$

We evaluate the integral

$$\int \frac{dx}{\sin xa}$$

by writing sin x = 2 sin x/2 cos x/2 = 2 tan x/2 cos x/2 and setting u = tan x/2 so that du sec$x$/2dx; the integral then becomes

$$\int \frac{dx}{\sin x} = \int \frac{du}{u} = \log\left|\tan\frac{x}{2}\right|$$

If we replace x by $x + \pi/2$, this formula becomes

$$\int \frac{dx}{\cos x} = \log\left|\tan\left(\frac{x}{2}+\frac{\pi}{4}\right)\right|$$

If we also apply the known trigonometrical formulae 2 cos$x$ = 1 + cos 2x and 2 sin x = 1 – cos 2x, the substitution u = 2x yields

$$\int \cos^2 dx = \frac{1}{2}(x+\sin x\cos x)$$

and $$\int \sin^2 xdx = \frac{1}{2}(x-\sin x\cos x)$$

By the substitution x = cos u, equivalent to u = arcos x or, more generally,

$x = a \cos u$ (a ⪢ 0), we can reduce

$$\int\sqrt{(1-x^2)}dx \text{ and } \int\sqrt{(a^2-x^2)}dx$$

respectively, to these formulae. We thus obtain

$$\int\sqrt{(a^2-x^2)}dx = -\frac{a^2}{2}\text{arccos}\frac{x}{a}+\frac{x}{2}\sqrt{(a^2-x^2)}$$

Similarly, the substitution $x = a \cosh u$ yields,

$$\int\sqrt{(x^2-a^2)}dx = -\frac{a^2}{2}\text{arcos}\,h\frac{x}{a}+\frac{x}{2}\sqrt{(x^2-a^2)}$$

and the substitution $x = a \sinh u$

$$\int\sqrt{(a^2+x^2)}\,dx = \frac{a^2}{2}\text{ar}\sin h\frac{x}{a}+\frac{x}{2}\sqrt{(a^2+x^2)}$$

The substitution $u = a/x$, $dx = -(a/u)\,du$ leads to

$$\int\frac{dx}{x\sqrt{(x^2-a^2)}} = -\frac{1}{a}\text{arcsin}\frac{a}{x}$$

$$\int\frac{dx}{x\sqrt{(x^2+a^2)}} = -\frac{1}{a}\text{ar}\sinh\frac{a}{x}$$

$$\int \frac{dx}{x\sqrt{\left(a^2 - x^2\right)}} = -\frac{1}{a}\operatorname{ar\,cosh}\frac{a}{x}$$

Finally, we consider the three integrals

$$\int \sin mx \sin nx dx, \int \sin mx \cos nx dx, \int \cos mx \cos nx dx$$

where m and n are positive integers. By the well-known trigonometrical formulae

$$\sin mx \sin nx = \frac{1}{2}\left\{\cos(m-n)x - \cos(m+n)x\right\}$$

$$\sin mx \cos nx = \frac{1}{2}\left\{\sin(m+n)x + \sin(m-n)x\right\}$$

$$\cos mx \cos nx = \frac{1}{2}\left\{\cos(m+n)x + \cos(m-n)x\right\}$$

we can divide each of these integrals into two parts. If we now use the substitutions $u = (m+n)x$ and $u = (m-n)x$, respectively, we obtain directly the system of formulae:

$$\int \sin mx \sin nx dx = \begin{cases} -\frac{1}{2}\left\{\frac{\cos(m+n)x}{m+n} + \frac{\cos(m-n)x}{m-n}\right\} \text{if } m \neq n \\ -\frac{1}{2}\left(x - \frac{\sin 2mx}{2m}\right) \text{if } m = n \end{cases}$$

$$\int \sin mx \cos nx dx = \begin{cases} -\frac{1}{2}\left\{\frac{\cos(m+n)x}{m+n} + \frac{\cos(m-n)x}{m-n}\right\} \text{if } m \neq n \\ -\frac{1}{2}\left(\frac{\cos 2mx}{2m}\right) \text{if } m = n \end{cases}$$

$$\int \cos mx \cos nx dx = \begin{cases} \frac{1}{2}\left\{\frac{\sin(m+n)x}{m+n} + \frac{\sin(m-n)x}{m-n}\right\} \text{if } m \neq n \\ \frac{1}{2}\left(\frac{\cos 2mx}{2m} + x\right) \text{if } m = n \end{cases}$$

In particular, if we now integrate from $-\pi$ to $+\pi$, we obtain from these formulae the extremely important relations

$$\int_{-\pi}^{+\pi} \sin mx \sin nx dx = \begin{cases} 0 \text{ if } m \neq n \\ \pi \text{ if } m = n \end{cases}$$

$$\int_{-\pi}^{+\pi} \sin mx \cos nx dx = 0$$

$$\int_{-\pi}^{+\pi} \cos mx \cos nx dx = \begin{cases} 0 \text{ if } m \neq n \\ \pi \text{ if } m = n \end{cases}$$

These are the orthogonality relations of the trigonometric functions which we shall meet again in Chap ix.

Integration by Parts

The second useful method for dealing with integration problems is given by the differentiation formula for a product:

$$(fg)' = f'g + fg'$$

If we write this formula as an integral formula, we obtain

$$f(x)g(x) = \int g(x)f'(x)dx + \int f(x)g'(x)dx$$

or

$$\int f(x)g'(x)dx = f(x)g(x) - \int g(x)f'(x)dx$$

This formula is referred to as the formula for integration by parts. The calculation of one integral is thereby reduced to the calculation of another integral. In fact, if we split up the integrand of an integral $\int w(x)dx$ into a product $\omega(x) = f(x)\,\phi(x)$ and find the indefinite integral

$$g(x) = \int \phi(x)dx$$

of the one factor $\phi(x)$, so that $\phi(x) = g'(x)$, then our formula reduces the integral

$$\int \omega(x)dx = \int f(x)\phi(x)dx = \int f(x)g'(x)dx$$

to the integral $\int g(x)f'(x)dx$, which in some cases is more readily evaluated than the original form. Since a given function $\omega(x)$, which occurs as an integrand, can be regarded as a product $f(x)\phi(x) = f(x)g'(x)$ in a great many different ways, this formula provides us with a very effective tool for the transformation of integrals.

Written as a formula for definite integration, the formula for integration by parts is

$$\int_a^b f(x)g'(x)dx = f(x)g(x)\Big|_a^b - \int_a^b g(x)f'(x)dx$$

$$= f(b)g(b) - f(a)g(a) - \int_a^b g(x)f'(x)dx$$

In fact, in order to obtain the formula for definite integration from the formula for indefinite integration, we only need replace, first of all, the variable appearing on both sides in the formula for the indefinite integral by the value $x = b$, then by the value $x = a$ and write down the difference of these two expressions.

A simple interpretation of this formula, at least with suitable restrictions on the functions involved, can be given. Suppose that $y = f(x)$ and $z = g(x)$ are monotonic, and that $f(x)=A$, $f(b)=B$, $g(a)=\alpha$, $g(b)=\beta$; we can then form the inverse of the first function and substitute it into the equation, thus obtaining z as a function of y.

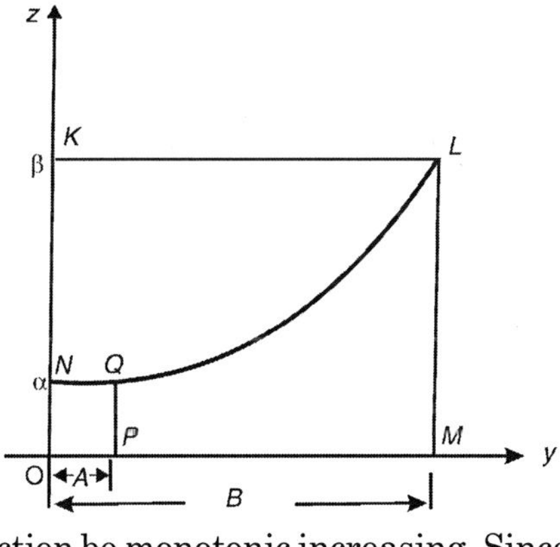

Let this function be monotonic increasing. Since $dy = f'(x)dx$ and $dz = g'(x)dx$, the formula for integration by parts can be rewritten

$$\int_A^B zdy + \int_\alpha^\beta ydz = B\beta - A\alpha$$

in agreement with the relation made clear by Figure,

$$\text{area } NQLK + \text{area } PMLQ = \text{area } OMLK - \text{area } OPQN$$

The following example may serve as a first illustration:

$$\int \log x dx = \int \log x.1.dx$$

We write the integrand in this way, in order to indicate that we intend to set $f(x) = \log x$ and $g'(x)=1$, so that we have $f'(x) = 1/x$ and $g(x) = x$. Our formula then becomes

$$\int \log x dx = x\log x - \int \frac{x}{x} dx = x\log x - x$$

This last expression is therefore the integral of the logarithm, as may be verified at once by differentiation.

Recurrence Formulae

In many cases, the integrand is not only a function of the independent variable, but also of an integral index n; on integrating by parts, instead of the value of the integral, we obtain another similar expression in which the index n has a smaller value. We thus arrive, after a number of steps, at an integral, which we can process by means of our table of integrals. Such a procedure is called a recurrence process.

The following examples illustrate this method: By repeated integration by parts, we can calculate the trigonometrical integrals

$$\int \cos^n x dx, \int \sin^n x dx, \int \sin^m x \cos^n x dx$$

provided that m and n are integers. In fact, we find that

$$\int \cos^n x dx = \cos^{n-1} x \sin x + (n-1)\int \cos^{n-2} x \sin^2 x dx$$

here we can rewrite the right-hand side in the form,

$$\cos^{n-1} x \sin x + (n-1)\int \cos^{n-2} x dx - (n-1)\int \cos^n x dx$$

and thus obtain the recurrence relation,

$$\int \cos^n x dx = \frac{1}{n}\cos^{n-1} x \sin x + \frac{n-1}{n}\int \cos^{n-2} x dx$$

This formula enables us to keep on lowering the index in the integrand until we finally arrive at the integral

$$\int \cos x dx = \sin x \text{ or } \int dx = x$$

according to whether n is odd or even. In a similar way, we obtain the analogous recurrence formulae

$$\int \sin^n x dx = -\frac{1}{n}\sin^{n-1} x \cos x + \frac{n-1}{n}\int \sin^{n-2} x dx$$

and

$$\int \sin^m x \cos^n x dx = \frac{\sin^{m+1} x \cos^{n-1} x}{m+n} + \frac{n-1}{m+n}\int \sin^m x cos^{n-2} x dx$$

In particular, these formulae enable us to compute the integrals

$$\int \sin^2 x dx = \frac{1}{2}(x - \sin x \cos)$$

and

$$\int \cos^2 x dx = \frac{1}{2}(x + \sin x \cos x)$$

which we have already obtained by the method of substitution. It need hardly be mentioned that the corresponding integrals for the hyperbolic functions can be calculated in exactly the same manner.

Further recurrence relations are obtained by the transformations:

$$\int (\log x)^m dx = x(\log x)^m - m\int (\log x)^{m-1} dx$$

$$\int x^m e^x dx = x^m e^x - m\int x^{m-1} e^x dx$$

$$\int x^m \sin x dx = -x^m \cos x + m\int x^{m-1} \cos x dx$$

$$\int x^m \cos x dx = x^m \sin x - m\int x^{m-1} \sin x dx$$

$$\int x^\alpha (\log x)^m dx = \frac{x^{\alpha+1}(\log x)^m}{a+1} - \frac{m}{a+1}\int x^a (\log x)^{m-1} dx \quad (a \neq -1)$$

Wallis' Product

The recurrence formula for the integral $\int \sin^n x dx$ leads on an elementary way to a most remarkable expression for the number π as an infinite product. We suppose that $n > 1$ and insertthe limits 0 and $\pi/2$ in the formula

$$\int \sin^n x dx = -\frac{1}{n}\sin^{n-1} x \cos x + \frac{n-1}{n}\int \sin^{n-2} x dx$$

obtaining

$$\int_0^{\pi/2} \sin^n x dx = \frac{n-1}{n}\int_0^{\pi/2} \sin^{n-2} x dx \text{ for n > 1}$$

Repeated application of the recurrence formula to the right-hand side yields for the cases $n = 2m$ and $n = 2m + 1$

$$\int_0^{\pi/2} \sin^{2m} x dx = \frac{2m-1}{2m}\cdot\frac{2m-3}{2m-2}\cdots\frac{1}{2}\cdot\int_0^{\pi/2} dx$$

$$\int_0^{\pi/2} \sin^{2m+1} x dx = \frac{2m}{2m+1}\cdot\frac{2m-2}{2m-1}\cdots\frac{2}{3}\cdot\int_0^{\pi/2} \sin x dx$$

whence

$$\int_0^{\pi/2} \sin^{2m} x dx = \frac{2m-1}{2m}\cdot\frac{2m-3}{2m-2}\cdots\frac{1}{2}\cdot\frac{\pi}{2}$$

$$\int_0^{\pi/2} \sin^{2m+1} x dx = \frac{2m}{2m+1}\cdot\frac{2m-2}{2m-1}\cdots\frac{2}{3}\cdot$$

Division now yields

$$\frac{\pi}{2} = \frac{2.2}{1.3}\cdot\frac{4.4}{3.5}\cdot\frac{6.6}{5.7}\cdots\frac{2m.2m}{(2m-1)\cdot(2m+1)}\frac{\int_0^{\pi/2} \sin^{2m} x dx}{\int_0^{\pi/2} \sin^{2m+1} x dx}$$

The quotient of the two integrals on the right hand side converges to 1 as m increases, as we recognize from the following considerations. In the interval $0 < x < \pi/2$, we find

$$0 < \sin^{2m+1} x \leq \sin^{2m} x \leq \sin^{2m-1} x$$

whence

$$0<\int_0^{\pi/2}\sin^{2m+1}xdx \leqq \int_0^{\pi/2}\sin^{2m}xdx \leqq \sin^{2m-1}xdx$$

If we here divide each term by $\int_0^{\pi/2}\sin^{2m+1}xdx$ and note that by the first formula, proved above,

$$\frac{\int_0^{\pi/2}\sin^{2m-1}xdx}{\int_0^{\pi/2}\sin^{2m+1}xdx} = \frac{2m+1}{2m}=1+\frac{1}{2m}$$

we find that

$$1\leqq\frac{\int_0^{\pi/2}\sin^{2m}xdx}{\int_0^{\pi/2}\sin^{2m+1}xdx}\leqq 1+\frac{1}{2m}$$

which yields the statement above, whence there applies the relation

$$\frac{\pi}{2} = \lim_{m\to\infty}\frac{2}{1}\frac{2}{3}\frac{4}{3}\frac{4}{5}\frac{6}{5}\frac{6}{7}\cdots\frac{2m}{2m-1}\frac{2m}{2m+1}$$

This product formula (due to Wallis), with its simple law of formation, presents a remarkable relation between the number π and the integers.

If we observe that,

$$\lim_{m\to\infty}\frac{2m}{2m+1} = 1$$

we can write and, if we take the square root and then multiply numerator and denominator by 2.4...,... (2m — 2), we find

$$\sqrt{\frac{\pi}{2}} = \lim_{m\to\infty}\frac{2.4\ldots(2m-2)}{3.5\ldots(2m-1)}\sqrt{2m} = \lim_{m\to\infty}\frac{2^2.4^2\ldots(2m-2)^2}{(2m-1)!}\sqrt{2m}$$

$$=\lim_{m\to\infty}\frac{2^2.4^2\ldots(2m)^2}{(2m)!}\frac{\sqrt{2m}}{2m}$$

Hence, we finally obtain

$$\lim_{m\to\infty}\frac{(m!)^2\,2^{2m}}{(2m)!\sqrt{m}} = \sqrt{\pi}$$

Integration of Rational Functions

The most important general class of functions, which are integrable in terms of elementary functions, are the rational functions

$$R(x) = \frac{f(x)}{g(x)}$$

where $f(x)$ and $g(x)$ are the polynomials:

$$f(x) = a_m x^m + a_{m-1}x^{m-1} + \ldots + a_0$$

$$g(x) = b_n x^n + a_{n-1}x^{n-1} + \ldots + b_0\,(b_n \neq 0)$$

We recall that every polynomial can be integrated at once and that the integral is itself a polynomial, whence we need only consider those rational functions for which the denominator is not a constant. Moreover, we can always assume that the degree of the numerator is less than the degree (n) of the denominator, because otherwise we can divide the polynomial $f(x)$ by the polynomial $g(x)$ and obtain a remainder of degree less than n; in other words, we can write $f(x)=q(x)g(x)+r(x)$, where $q(x)$ and $r(x)$ are also polynomials and $r(x)$ is of degree less than n. The integration of $f(x)/g(x)$ is then reduced to the integration of the polynomial $q(x)$ and of the proper fraction $r(x)/g(x)$. Moreover, we note that the $f(x)/g(x)$ can be represented as the sum of the functions $a_\nu x^\nu/g(x)$, so that we need only consider integrands of the form $x^\nu/g(x)$.

Chaper 16

Improper Integrals

Infinite Interval of Integration

Another important extension of the concept of integral occurs when one of the limits of integration as infinite. In order to make this extension precise, we introduce the notation: If the integral

$$\int_a^A f(x)\,dx$$

where a is fixed, tends to a definite limit when A increases positively beyond all bounds, we denote the limit by

$$\int_a^\infty f(x)\,dx$$

and call it the integral of the function $f(x)$ from a to ∞. Of course, such an integral does not necessarily exist or, as we often will say, converge.

Simple examples of the various possibilities are again given by the function $f(x)=1/x^{\alpha}$:

$$\int_1^A \frac{dx}{x^\alpha} = \frac{1}{1-\alpha}\left(A^{1-\alpha}-1\right)$$

We see here that, if we again exclude the case $\alpha = 1$, the integral to infinity exists for the case $\alpha > 1$, and, in fact,

$$\int_1^\infty \frac{dx}{x^\alpha} = \frac{1}{\alpha-1}$$

In contrast, when $\alpha < 1$, the integral no longer exists. In the case $\alpha = 1$, the integral again clearly fails to exist, since log x tends to infinity with x. Hence we see that, as regards integration over an infinite interval, the functions $1/x^{\alpha}$ do not behave in the same way as for integration up to the origin.

This statement also is made plausible by a glance at fiigure. In, we see that, the larger is α, the more slowly do the curves draw in towards the x-axis when x is large, so that we can readily assume that the area under consideration tends to a definite limit for sufficiently large values of α.. The following criterion for the existence of an integral with an infinite limit is often useful. We again assume that for sufficiently large valued of x, say, for $x<\alpha$, the integrand has always the same sign, which, without loss of generality, we can choose to be positive.

Then we have the statement: The integral $\int_a^{\infty} f(x)dx$ converges if the function $f(x)$ vanishes at infinity to a higher order than the first, that is, if there is a number $\nu > 1$ such that for all values of x, no matter how large, the relation $0<f(x) \} M/x^{\nu}$ is true, where M is a fixed number independent of x. Again, the integral diverges, if the function remains positive and vanishes at infinity to an order not higher than the first, that is, if there is a fixed number $N > 0$ such that $xf(x)= N$.

The proof of these criteria, which run exactly parallel to the previous argument, is left to the reader. A very simple example is the integral $\int_a^{\infty} \frac{1}{x^2} dx (a > 0)$. The integrand vanishes at infinity to the second order. As a matter of fact, we see at once that the integral does converge, because

$$\int_a^{A} \frac{1}{x^2} dx = \frac{1}{a} - \frac{1}{A} \text{ whence}$$

$$\int_a^{\infty} \frac{1}{x^2} dx = \frac{1}{a}$$

Another equally simple example is

$$\int_0^{\infty} \frac{1}{1+x^2} dx = \lim_{A \to \infty} (\arctan A - \arctan 0) = \frac{\pi}{2}$$

The Gamma Function

Another example of particular importance in analysis is offered by the socalled Γ–function

$$\Gamma(n) = \int_0^\infty e^{-x}x^{n-1}dx \ (n>0)$$

Here too the criterion of convergence is satisfied; for example, if we choose ν=2, we have since the exponential function e^x tends to zero to a higher order than any power $1/x^m$ $(m > 0)$. This gamma function, which we can think of as a function of the number n (not necessarily an integer), satisfies a remarkable relation, which we can arrive at in the following way by integration by parts: To begin with, we have

$$\int e^{-x}x^{n-1}dx = -e^{-x}x^{n-1} + (n-1)\int e^{-x}x^{n-2}dx$$

If we take this formula between the limits 0 and A and then let A increase beyond all bounds, we immediately obtain

$$\Gamma(n) = (n-1)\int_0^\infty e^{-\alpha}x^{n-2}dx = (n-1)\Gamma(n-1)$$

and by this recurrence formula, provided μ is an integer and $0 < \mu < n$,

$$\Gamma(n) = (n-1)(n-2)...(n-\mu)\int_0^\infty e^{-\alpha}x^{n-\mu-1}dx$$

In particular, if n is a positive integer, we have

$$\Gamma(n) - (n-1)(n-2)...3.2.1\int_0^\infty c^{-\alpha}dx$$

and, since

$$\int_0^\infty e^{-\alpha}dx = 1$$

it follows finally that $\Gamma(n) = (n-1)(n-2)...2.1 = (n-1)!$.

This expression for a factorial by an integral is of importanoe in many applications. The integrals,

$$\int_0^\infty e^{-x^2}dx, \int_0^\infty x^n e^{-\alpha^2}dx$$

also converge, as we may easily verify by means of our criterion.

The Dirichlet Integral

A convergent integral, important in many applications, the convergence of which does not follow directly from our criterion and which is a simple case of a type investigated by Dirichlet, is

$$I = \int_0^\infty \frac{\sin x}{x} dx$$

This integral is easily seen to be convergent, if the upper limit is finite, because $\sin x/x \to 1$ as $x \to 0$. Its convergence in the infinite interval is due to the periodic change of sign of the integrand, which causes the contributions to the integral from neighbouring intervals of length π almost to cancel each another. In order to utilize this fact, we write the expression

$$D_{AB} = \int_A^B \frac{\sin}{x} dx$$

in the form

$$D_{AB} = \int_A^{A+\pi} \frac{\sin x}{x} dx - \int_B^{B+\pi} \frac{\sin x}{x} dx + \int_{A+\pi}^{B+\pi} \frac{\sin t}{t} dt$$

introduce in the last of the three integrals the new variable $x = t - p$, whence $\sin t = -\sin x$, and obtain

$$D_{AB} = \int_A^{A+\pi} \frac{\sin x}{x} dx - \int_B^{B+\pi} \frac{\sin x}{x} dx - \int_A^B \frac{\sin x}{x+\pi} dx$$

Addition of this expression to the original one for D_{AB} yields

$$2D_{AB} = \int_A^{A+\pi} \frac{\sin x}{x} dx - \int_B^{B+\pi} \frac{\sin x}{x} dx + \pi \int_A^B \frac{\sin x}{x(x+\pi)} dx$$

Hence, if we assume that $B > A > 0$, it follows that

$$|2D_{AB}| < \frac{2\pi}{A} + \pi \int_A^B \frac{dx}{x^2}$$

in fact, while observing that

$$-\frac{1}{x} \leq \frac{\sin x}{x} \leq \frac{1}{x}$$

and

$$-\frac{1}{x^2} \leqq \frac{\sin x}{x(x+\pi)} \leqq \frac{1}{x^2}$$

for positive values of x. The integral on the right hand side converges, by our criterion and our formula shows that $|D_{AB}|$ ⟩ 0 as both A and B tend to infinity. Now,

$$|D_{OB} - D_{OA}| = |D_{AB}|$$

and it follows from Cauchy's convergence test that D_{0B} tends to a definite limit as B. In other words, the integral I exists.

Substitution

It is obvioua that all rules for the substitution of new variables, etc. remains valid for convergent improper integrals. As an example, in order to calculate $\int_0^{\infty} xe^{-x^2} dx$, we introduce the new variable $u = x$ and obtain

$$\int_0^{\infty} xe^{-x^2} dx = \frac{1}{2}\int_0^{\infty} e^{-u} du = \lim_{A\to\infty} \frac{1}{2}\left(1 - e^{-A}\right) = \frac{1}{2}$$

Other examples of the use of substitution in the investigation of improper integrals are the Fresnel integrals, which occur in the theory of diffraction of light:

$$F_1 = \int_0^{\infty} \sin\left(x^2\right) dx, F_2 = \int_0^{\infty} \cos\left(x^2\right) dx$$

The substitution $x = u$ yields

$$F_1 = \frac{1}{2}\int_0^{\infty} \frac{\sin u}{\sqrt{u}} du, F_2 = \frac{1}{2}\int_0^{\infty} \frac{\cos u}{\sqrt{u}} du$$

Integrating by parts, we have

$$\int_A^B \frac{\sin u}{\sqrt{u}} du = \frac{\cos A}{\sqrt{A}} - \frac{\cos B}{\sqrt{B}} - \frac{1}{2}\int_A^B \frac{\cos u}{u^{3/2}} du$$

The integral also tends to 0. Hence, by the same argument as for the Dirichlet integral, we see that the integral F_1 converges. The convergence of the integral F_2 is proved in exactly the same way.

These Fresnel integrals show that an improper integral may exist even although the integrand does not tend to zero as x. In fact, an improper integral can exist even when the integrand is unbounded, as is shown by the example

$$\int_0^\infty 2u\cos\left(u^4\right)du$$

When $u^4 = n\pi$, i.e. when $u = \sqrt[4]{n\pi}$, $n = 0, 1, 2, \ldots$, the integrand becomes $2\sqrt[4]{n\pi}\cos n\pi = \pm 2\sqrt[4]{n\pi}$, so that the integrand is unbounded. However, the substitution $u = x$ reduces the integral to

$$\int_0^\infty \cos\left(x^2\right)dx$$

which we have just shown to be convergent.

By means of a substitution, an improper integral may often be transformed into a proper one. For example, the transformation $x = \sin u$ yields

$$\int_0^1 \frac{dx}{\sqrt{\left(1-x^2\right)}} = \int_0^{\pi/2} du = \frac{\pi}{2}$$

On the other hand, integrals of continuous functions may be transformed into improper integrals; this oocurs if the transformation $u = \phi(x)$ is such that at the end of the interval of integration the derivative $\phi'(x)$ vanishes, so that dx/du is infinite.

The Second Mean Value Theorem of the Integral Calculus

The method of integration by parts yields us an easy way for proving n important theorem on the estimation of integrals, usually called the second mean value theorem of the integral calculus.

Let the function $\phi(x)$ be monotonic and continuous in the interval $a < x < b$, the derivative $\phi'(x)$ be continuous and moreover $f(x)$ be an arbitrary function which is continuous in the same interval. Then one has the second mean value theorem of the integral calculus:

There exists a number ξ such that $a < \xi < b$ for which

$$\int_a^b f(x)\phi(x)dx = \phi(a)\int_a^\xi f(x)dx + \phi(b)\int_\xi^b f(x)dx$$

We prove this by noting first that we can assume that $\phi(b) = 0$; in fact, replacement of $\phi(x)$ by $\phi(x) - \phi(b)$ changes both sides of the equation by the same amount and yields a function which vanishes at $x = b$.

Moreover, we can assume that $\phi(a) > 0$; in fact, if $\phi(a) < 0$, we need only replace $\phi(x)$ by $-\phi(x)$, which changes the sign of both sides of the equation. (The case $\phi(a) = 0$ is trivial; in fact, if both $\phi(a)$ and $\phi(b)$ vanish, $\phi(x)$ must be identically zero and our equation becomes $0 = 0$.)

Hence, we must only prove that, if $\phi(x)$ is continuous and monotonic decreasing and $\phi(b) = 0$, then

$$\int_a^b f(x)\phi(x)dx = \phi(a)\int_a^\xi f(x)dx$$

We now set $F(x) = \int_a^x f(x)dx$ and apply the formula for integration by parts to the left hand side of the last equation; we then have

$$\int_a^x f(x)\phi(x)dx = F(x)\phi(x)\Big|_a^b + \int_a^b F(x)\{-\phi'(x)\}dx$$

The integrated part vanishes, since $F(a)$ and $\phi(b)$ are zero. The expression $-\phi'(a)$ is everywhere positive, so that we can apply the first mean value theorem of the integral calculus.

We thus find that the integral on the right hand side has the value

$$F(\xi)\int_a^b \{-\phi'(x)\}dx, a \le \xi \le b$$

However,

$$F(\xi) = \int_a^\xi f(x)dx$$

and

$$\int_a^b \{-\phi'(x)\}dx = \phi(a) - \phi(b) = \phi(a)$$

and our theorem is established. This theorem can be extended (although we shall not carry out the proof) to more general classes of functions. The theorem remains true for all continuous monotonic functions $\phi(x)$ whether they have derivatives or not. In fact, it is true for any discontinuous monotonic function for which we are in a position to integrate $f(x)\phi(x)$.

Numerical Methods

Anyone who has to use analysis as an instrument for investigating physical or technical phenomena is faced by the question whether and how the theory can be adapted to yield useful practical methods for actual numerical calculations. Yet, even from the point of view of the theoretical person, who only desires to recognize the link between natural phenomena and not to conquer them, these questions are of no trifling interest. Here we can only discuss some particularly important points which are more or less closely linked to the preceding ideas. We wish to direct special attention to the fundamental fact that the meaning of an approximate calculation is not precise unless it is accompanied by a definite knowledge of the degree of accuracy attained.

Numerical Integration

We have seen that even relatively simple functions cannot be integrated in terms of elementary functions and that it is futile to make this unattainable goal the aim of the integral calculus. On the other hand, the definite integral of a continuous function does exist and this fact raises the problem of finding methods for calculating it numerically. Here we shall discuss the simplest and most obvious of these methods with the aid of geometrical intuition and thenconsider error estimation.

Our objective is to calculate the integral

$$I = \int_a^b f(x)dx,$$

where $a < b$. We imagine the interval of integration to have been subdivided into n equal parts of length $h = (b - a)/n$ and denote the points of subdivision by $x_1=a$, $x_2=a+h$, ..., $x_n=b$, the values of the function at the points of subdivision by $f_0, f_1, \ldots, f_n$ and, similarly, the values of the function at the midpoints of the intervals by $f_{1/2}$, $f_{3/2}$, ..., $f_{(2n-1)/2}$. We interpret our integral as an area and cut up the region under the curve into strips of width h in the usual manner. We must now obtain an approximation for each such strip of surface, that is, for the integral

$$I_v = \int_{x_y}^{x_y+h} f(x)dx,$$

The Rectangle Rule: The crudest and most obvious method of approximating the integral I is directly linked to the definition of the integral; we replace the area of the strip I_n by the rectangle of area $f_n h$ and obtain for the integral the approximate expression

$$I \approx h(f_0 + f_1 + ... + f_{n-1})$$

From here on, the symbol ≈ means is approximately equal to.

$$x_v, x_{v+1} = x_v + h, \text{ and } x_{v+2} = x_v + 2h$$

The Trapezoid and Tangent Formulae: We obtain a closer approximation with no greater trouble if we do not replace the area of the strip I_v, as above, by a rectangular area, but by the trapezoid of area $(f_v + f_{v+1})h/2$. For the whole integral, this process yields the approximate expression

$$I \approx h(f_1 + f_2 + ... + f_{n-1}) + \frac{h}{2}(f_0 + f_n)$$

(trapezoid formula), since, when the areas of the trapezoids are added, each value of the function except the first and the last occurs twice. As a rule, the approximation becomes even better if, instead of choosing the trapezoid under the chord AB as an approximation to the area of I, we select the trapezoid under the tangent to the curve at the point with the abscissa $x = x_v + h/2$. The area of this trapezoid is simply $hf_{v+½}$, so that the approximation for the entire integral is

$$I \approx h(f_{1/2} + f_{3/2} + ... + f_{2n-1/2}),$$

which is called the tangent formula.

Simpson's Rule: By means of Simpson's rule, we arrive with very little more trouble at a numerical result which is generally much more exact. This rule depends on estimating the area I_v

$+ I_{v+1}$ of the double strip between the abscissae $x = x_v$ and $x = x_v + 2 = x_{v+2}$ by considering the upper boundary to be no longer a straight line but a parabola - in order to be specific, that parabola which passes through the three points of the curve with the abscissae x_v, $x_{v+1} = x_v + h$, *and* $x_{v+2} = x_v + 2h$. The equation of this parabola is,

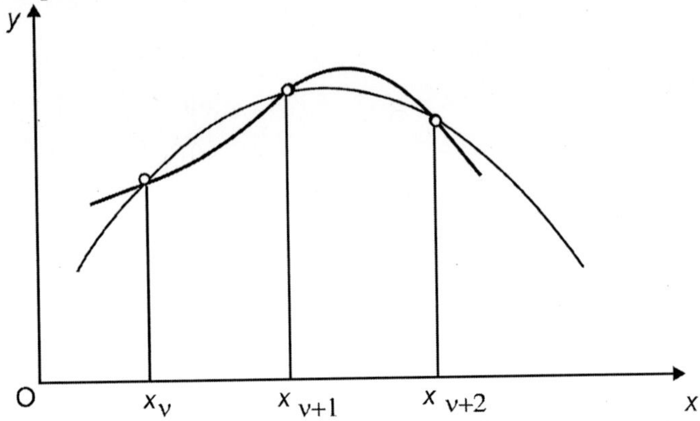

***Fig.** Simpson's rule*

The student should verify by direct substitution that for the three values of x in questiŏn this eqnation gives the proper values of y, i.e., f_n, f_{n+1} and f_{n+2}, respectively. If we integrate this second degree polynomial between the x_n and $x_n + 2h$, we obtain, after a brief calculation, for the area under the parabola

$$\int_{xv}^{xv+2h} y\,dx = \begin{array}{l} 2hf_v + 2h\left(f_{v+1} - f_v\right) \\ + \dfrac{1}{2}\left(\dfrac{8}{3}h - 2h\right)\left(f_{v+2} - 2f_{v+1} + f_v\right) \end{array}$$

$$= \frac{h}{3}\left(f_v + 4f_{v+1} + f_{v+2}\right)$$

This represents the required approximation to the area of our strip $I_n + I_{n+1}$. If we now assume that $n = 2m$, i.e., that n is an even number, we obtain, by addition of the areas of such strips, Simpson's rule,

$$I \approx \frac{4h}{3}\left(f_1 + f_3 + + \ldots f_{2m-1}\right) + \frac{2h}{3}\left(f_2 + f_4 + \ldots + f_{2m-2}\right) + \frac{h}{3}\left(f_0 + f_{2m}\right)$$

Estimation of the Error: It is easy to obtain an estimate of the error for each of our methods of integration, if the derivatives of the function $f(x)$ are known throughout the interval of the integration.

We take M_1, M_2,... as the upper bounds of the absolute values of the first, second, derivative, respectively, i.e., we assume that throughout the interval $|f^{(\nu)}(x)| < M_\nu$. Then the estimates are:

For the rectangle rule:

$$|I_\nu - hf_\nu| < \frac{1}{2}M_1h^2 or \left|I - h\sum_{\nu=0}^{n-1} f_\nu\right| < \frac{1}{2}M_1nh^2 = \frac{1}{2}M_1(b-a)h$$

For the tangent rule:

$$\left|I_\nu - hf_{\nu+\frac{1}{2}}\right| < \frac{M}{24}h^3 \; or \left|I - h\sum_{\nu=0}^{n-1} f_{\nu+\frac{1}{2}}\right| < \frac{M_2}{24}(b-a)h^2$$

For the trapezoid rule:

$$\left|I_\nu - \frac{h}{2}(f_\nu + f_{\nu+1})\right| < \frac{M_2}{12}h^3$$

For Simpson's rule:

$$\left|I_\nu + I_{\nu+1} - \frac{h}{3}(f_\nu + 4f_{\nu+1} + f_{\nu+2})\right| < \frac{M_4}{90}h^5$$

The last two estimates also lead to estimates for the entire integral *1*.

We see that Simpson's rule has an error of much higher order in the small quantity h than the other rules, so that when M_4 is not too large, it is very advantageous for practical calculations. In order to avoid tiring the reader with the details of the proofs of these estimates, which are fundamentally quite simple, we shall restrict ourselves to the proof for the tangent formula. For this purpose, we expand the function $f(x)$ in the $(n + 1)$-th strip by Taylor's theorem:

$$f(x) = f_{\nu+\frac{1}{2}} + \left(x - x_\nu - \frac{h}{2}\right)f'\left(x_\nu + \frac{h}{2}\right) + \frac{1}{2}\left(x - x_\nu - \frac{h}{2}\right)^2 f''(\xi),$$

where ξ is a certain intermediate value in the strip. If we integrate the right hand side over the interval $x_n < x < x_n + h$, the integral of the middle term is zero. Since, as is easily verified,

$$\frac{1}{2}\int_{x_\nu}^{x_\nu+h}\left(x - x_\nu - \frac{h}{2}\right)^2 dx = \frac{h^3}{24},$$

it follows immediately that

$$\left|\int_{x_\nu}^{x_\nu+h} f(x)dx - hf_{\nu+\frac{1}{2}}\right| < M_2\frac{h^3}{24},$$

which proves the above assertion.

Mean Value theorem and of Taylor's Theorem

The Calculus of Errors: We now come to quite a different type of numerical calculations. These are applications of the mean value theorem or, more generally, of Taylor's theorem with remainder or, finally, of the infinite Taylor series.

As an application which, although simple, is quite important in practice, we shall consider the calculus of errors. This rests on the idea - which lies at the root of the entire differential calculus - that a function $f(x)$, which is differentiable a sufficient number of times, can be represented in the neighbourhood of a point by a linear function with an error of order less than the first, by a quadratic function with an error of order less than the second, etc. Consider the linear approximation to a function $y=f(x)$. If $y+\Delta y=f(x+\Delta x)=f(x+h)$, we find by Taylor's theorem

$$\Delta y = hf'(x) + \frac{h^2}{2}f''(\xi)$$

where $\xi = x + \theta h$ $(0 < \theta < 1)$ is an intermediate value, which need not be known more precisely. If $h = \Delta x$ is small, we obtain as a practical approximation $\Delta y \approx hf'(x)$.

In other words, we replace the difference quotient by the derivative to which it is approximately equal and the increment of y by the approximately equal linear expression in h.

We use this fairly obvious fact for practical purposes in the following way. Let two physical quantities x and y be linked by the relation $y = f(x)$. There arises then the question how an inaccuracy in the measurement of x affects the determination of y. If we happen to use instead of the true value of x the inaccurate value $x + h$, then the corresponding value of y differs from the true value $y = f(x)$ by $\Delta y = f(x + h) - f(x)$, whence the error is given approximately by the above relation.

Evaluation of π: Gregory's series, also called Leibnitz' series,

$$\frac{\pi}{4} = 1 - \frac{1}{3} + \frac{1}{5} - \frac{1}{7} + - \ldots,$$

using the series for artan, is not suitable for the calculation of π due to the slowness of its convergence. However, we may calculate π with comparative ease by the artifice: From the addition theorem for the tangent

$$\tan(\alpha + \beta) = \frac{\tan\alpha + \tan\beta}{1 - \tan\alpha\ \tan\beta},$$

changing to the inverse functions $\alpha = \text{artan}\ u$, $\beta = \text{artan}\ v$, we obtain

$$\text{arc tan}u + \text{arc tan}v = \text{arc tan}\left(\frac{u+v}{1-uv}\right)$$

If we now choose u and v in such a manner that $(u + v)/(1 - uv) = 1$, we obtain the value $\pi/4$ on the right hand side and, if u and v are small numbers, we can easily compute the left hand side by means of known series. For example, if we set $u = 1/2$, $v = 1/3$, as Euler did, we obtain $\frac{\pi}{4} = \text{arc tan}\frac{1}{2} + \text{arc tan}\frac{1}{3}$. Moreover, if we note that $\left(\frac{1}{3} + \frac{1}{7}\right) \div \left(1 - \frac{1}{21}\right) = \frac{1}{2}$

we have $\text{arc tan}\frac{1}{2} = \text{arc tan}\frac{1}{3} + \text{arc tan}\frac{1}{7}$, so that $\frac{\pi}{4} = 2\ \text{arc tan}\frac{1}{7} + \text{arc tan}$.

Using this formula, Georg Vega (1756 - 1802) calculated the number π to 140 places. Moreover, using the equation,

$$\left(\frac{1}{5}+\frac{1}{8}\right)\div\left(1-\frac{1}{40}\right)=\frac{1}{3}$$

we obtain arc $\tan\frac{1}{3}=$ arc $\tan\frac{1}{5}+$arc $\tan\frac{1}{8}$ or $\frac{\pi}{4}=2$ arc $\tan\frac{1}{5}+$arc $\tan\frac{1}{7}+2$ arc $\tan\frac{1}{8}$. This expansion is extremely useful for the computation of π by means of the series artan$x$ $=x-\frac{x^3}{3}+\frac{x^5}{5}-+\ldots$; in fact, if we substitute for x the values 1/5, 1/7 or 1/8, we obtain with just a few terms a high degree of accuracy, since the terms diminish rapidly. However, we can perform the calculation even more conveniently by basing it on the formula

$$\frac{\pi}{4}=\text{arc tan}\frac{120}{119}-\text{arc tan}\frac{1}{239}=4\text{ arc tan}\frac{1}{5}-\text{arc tan}\frac{1}{239}$$

obtained by similar arguments as those above. Calculation of Logarithms: For the computation of logarithms, we transform the series,

$$\frac{1}{2}\log\frac{1+x}{1-x}=x+\frac{x^3}{3}+\frac{x^5}{5}+\ldots(|x|<1);$$

where $0<x<1$, by the substitution

$$\frac{1+x}{1-x}=\frac{p^2}{p^2-1},\ \text{x}=\frac{1}{1p^2-1}$$

into the series

$$\log p=\frac{1}{2}\log(p-1)+\frac{1}{2}\log(p+1)+\frac{1}{2p^2-1}+\frac{1}{3(2p^2-1)^3}+\ldots,$$

where $2p-1>1$, thatis, $p>1$. f p is an integer and $p+1$ can be resolved into smaller integral factors, this last series expresses the logarithm of p in terms of the logarithms of smaller integers plus a series, the terms of which diminish very rapidly and the sum of which can therefore be calculated accurately enough fromonly a fe terms. Hence this series allows us to compute thelogarithmsof any prme number and hence tha of any nuber, provided we have already calculated the value of log 2.

The accuracy of this determination of log p can be estimated more easily by mans of thegeometricseries than from he genera remainder formula.

In fact, we have for the remainder R_n of the series, i.e., the sum of all the terms following the term,

$$\frac{1}{n\left(2p^2-1\right)^n}$$

$$R_n \frac{1}{(n+2)\left(2p^2-1\right)^{n+2}}\left(1+\frac{1}{\left(2p^2-1\right)^2}+\frac{1}{\left(2p^2-1\right)^4}+\ldots\right)$$

$$=\frac{1}{(n+2)\left(2n^2-1\right)^n}\cdot\frac{1}{\left(2p^2-1\right)^2-1}$$

and this formula yields immediately the required error estimate. For example, let us compute $\log_e 7$, using the first four terms of the series. We have

P = 7, 2P^2 – 1 = 97,

$$\log 7 = 2\log 2 + \frac{1}{2}\log 3 + \frac{1}{97} + \frac{1}{3.97^3} \text{ I} \ldots, \frac{1}{97} \approx$$

$.01030928, \approx 0.00000037,$

$2 \log 2 \approx 138629436, \frac{1}{2}$

$\text{Log}3 \approx 0.54930614,$

whence $\log_e 7 \approx 1.945,910,15.$

Estimation of the error yields,

$$R_n < \frac{1}{5.94^3} \times \frac{1}{97^3-1} < \frac{1}{36\times 10^3}$$

However, we must note that each of the four numbers which we have added is only given to within an error, so that the last place in the above value of $\log_e 7$ might be wrong by 2. However, as a matter of fact, the last place is also correct.

Continuity of Functions

Differentiability and Continuity of Functions

It is useful to point out that, if we know that a function can be differentiated, we need not give any special proof of its continuity.

If a function is differentiable, then it is necessarily continuous. In fact, if the difference quotient $\frac{f(x+h)-f(x)}{h}$ approaches a definite limit as h tends to zero, the numerator of the fraction, that is, $f(x + h) - f(x)$, must tend to zero with h, and this fact expresses the continuity of the function $f(x)$ at the point x.

However, the converse of this is definitely false; it is not true that every continuous function has a derivative at every point. The simplest example, which disproves this assumption, is the function $f(x) = |x|$, i.e., $f(x) = -x$ for $x < 0$ and $f(x) = x$ for $x > 0$; its graph. At the point $x = 0$, this function is continuous, but has no derivative. The limit of $[f(x + h) - f(x)]/h$ is equal to 1 if h tends to 0 through positive values and equal to - 1 if h tends to zero through negative values; if we do not restrict the sign of h, There does not exist a limit.

We say that the function has different right hand and and left-hand derivatives at the point x, where by we mean the limiting values of $[f(x + h) - f(x)]/h$ as h approaches 0 through positive values only and negative values only, respectively. Thus, the differentiability of a function requires not merely that the right-hand and left-hand derivatives exist, but that they are equal. Geometrically, the inequality of the two derivatives means that the curve has a sharp corner.

Bibliography

Apostol, Tom M.: *Calculus: One-Variable Calculus with an Introduction to Linear Algebra,* Wiley, NY, 2007.

Boole, George: *A Treatise On The Calculus of Finite Differences*, Macmillan and Company, New York, 2010.

Boyer, Carl B.: *The history of the Calculus and its Conceptual Development*, Dover Publications, New York, 1999.

Boyer, Carl B.: *The history of the Calculus and its Conceptual Development*, Dover Publications, New York, 2009.

Courant, Richard: *Differential and Integral Calculus. Vol. II*, Wiley Classics Library, New York: John Wiley & Sons, 2018.

Courant, Richard: *Differential and Integral Calculus.* Wiley Classics Library, New York: John Wiley & Sons, 1997.

Courant, Richard; John, Fritz: *Introduction to Calculus and Analysis Volume 1*, Classics in Mathematics, Berlin, New York:Springer-Verlag, 2019.

Deborah, Hughes-Hallett: *Multivariable Calculus*, Wiley, New York, 2023.

Eisenbud, David; Harris, Joe: *The Geometry of Schemes*, Springer-Verlag, 2018.

Gerald, B. Folland: *Real Analysis: Modern Techniques and Their Applications,* John Wiley & Sons, NY, 2014.

Goursat, Édouard: *A Course in Mathematical Analysis: Vol 1: Derivatives and Differentials, Definite Integrals, Expansion in Series, Applications to Geometry*, Dover Publications, New York, 2009.

Hardy, Godfrey Harold: *A Course of Pure Mathematics*, Cambridge University Press, 1998.

Hildebrand, F. B.: *Finite-Difference Equations and Simulations*, Prentice-Hall, Englewood Cliffs, New Jersey, 2008.

Hille, Einar; Phillips, Ralph S.: *Functional Analysis and Semi-groups*, American Mathematical Society, Providence, R.I., 2014.

Jackson, J.D.: *Classical Electrodynamics*, New York: John Wiley & Sons, 2018.

Kline, Morris: *Calculus: An intuitive and physical approach*, John Wiley and Sons, NY, 1997.

Kock, Anders: *Synthetic Differential Geometry,* Cambridge University Press, Cambridge, 2016.

Larson, Ron; Hostetler, Robert P.; and Edwards, Bruce H.: *Calculus of a Single Variable: Early Transcendental Functions*, Houghton Mifflin Company, 2013.

Lieb, Elliott; Loss, Michael: *Analysis*, AMS Chelsea, 2017.

Marsden, J.E.: Vector Calculus, W. H. Freeman & Company, 1996.

Michael J. Crowe: *A History of Vector Analysis: The Evolution of the Idea of a Vectorial System*, Dover Publications; New York, 2007.

Morris, Kline: *Mathematical thought from Ancient to Modern Times,* Oxford University Press, 2010.

Phillips, Ralph S. and Hille, Einar: *Functional Analysis and Semi-groups*, American Mathematical Society, Providence, R.I., 2004.

Richtmeyer, D. and Morton, K.W.: *Difference Methods for Initial Value Problems*, Wiley, New York, 1997.

Schey, H. M.: *Div Grad Curl and all that: An Informal Text on Vector Calculus,* W. W. Norton & Company, UK, 2015.

Sutherland, W. A.: *Introduction to Metric and Topological Spaces*, Oxford University Press, Oxford, 2015.

Index

❑❑❑